Engineering Mechanics

Engineering Mechanics

Nayan Rana

RANDOM PUBLICATIONS
NEW DELHI (INDIA)

Engineering Mechanics

ISBN 978-93-5111-425-3

Published in 2014 in India by

RANDOM PUBLICATIONS

4376-A/4B, Gali Murari Lal, Ansari Road
New Delhi-110 002
Phone : +91-11-43580356, +91-11-23289044
e-mail: randomexports@gmail.com, sales@randompublications.com, info@randompublications.com

Reprinted 2019

Type Setting by: Friends Media, Delhi-110089
Digitally Printed at: Replika Press Pvt. Ltd.

Preface

Engineering mechanics is the application of mechanics to solve problems involving common engineering elements. It is an important sub-discipline of mechanical engineering, and its study is the in-depth study of some of the core modules of a mechanical engineering course - mechanics, dynamics etc. With roots in physics and mathematics, Engineering Mechanics is the basis of all the mechanical sciences like, civil engineering, materials science and engineering, mechanical engineering and aeronautical and aerospace engineering. Engineering Mechanics provides the "building blocks" of statics, dynamics, strength of materials, and fluid dynamics. Engineering mechanics is the discipline devoted to the solution of mechanics problems through the integrated application of mathematical, scientific, and engineering principles. Special emphasis is placed on the physical principles underlying modern engineering design.

Engineering mechanics apply Isaac Newton's laws of motion in describing the movement of objects. The objects can be machines, fluids, solid materials or animals including humans. Stephen Timoshenko is credited with first describing the movement of materials and is said to be the father of modern engineering mechanics. Engineering mechanics not only includes the elastic movement of materials in which they remain in their normal states, but also on the plastic deformation of materials in which they are permanently altered. Everything from individual particles to the tallest buildings to a rushing river experiences forces and torques. Each object—liquid or solid, large or small— responds differently under different conditions. Engineering mechanics examines these external forces by studying statics, dynamics, materials strength, elasticity, viscoelasticity and fluid dynamics. As a bridge between theory and application, engineering mechanics is used formulate new ideas and theories, discover and interpret phenomena, and develop experimental and computational tools.

This is a comprehensive book meeting complete requirements of Engineering Mechanics course of undergraduate syllabus, as it focuses on basic concepts of Engineering Mechanics for providing the fundamental knowledge required for understanding advanced subjects based on mechanics.

I thank all members of my team who have helped in the preparation of the book. My special thanks go to "Random Publication" who have published the book.

—Nayan Rana

Contents

1

Introduction

Dynamics is a branch of physics (specifically classical mechanics) concerned with the study of forces and torques and their effect on motion, as opposed to *kinematics*, which studies the motion of objects without reference to its causes.

Generally speaking, researchers involved in dynamics study how a physical system might develop or alter over time and study the causes of those changes. In addition, Isaac Newton established the undergirding physical laws which govern dynamics in physics. By studying his system of mechanics, dynamics can be understood. In particular, dynamics is mostly related to Newton's second law of motion. However, all three laws of motion are taken into consideration, because these are interrelated in any given observation or experiment.

The study of dynamics falls under two categories: linear and rotational. Linear dynamics pertains to objects moving in a line and involves such quantities as force, mass/inertia, displacement (in units of distance), velocity (distance per unit time), acceleration (distance per unit of time squared) and momentum (mass times unit of velocity). Rotational dynamics pertains to objects that are rotating or moving in a curved path and involves such quantities as torque, moment of inertia/rotational inertia, angular displacement (in radians or less often, degrees), angular velocity (radians per unit time), angular acceleration (radians per unit of time squared) and angular momentum (moment of inertia times unit of angular velocity). Very often, objects exhibit linear and rotational motion.

For classical electromagnetism, it is Maxwell's equations that describe the dynamics. And the dynamics of classical systems involving both mechanics and electromagnetism are described by the combination of Newton's laws, Maxwell's equations, and the Lorentz force.

Force

In physics, a force is any influence that causes an object to undergo a certain change, either concerning its movement, direction, or geometrical construction. In other words, a force can cause an object with mass to change its velocity (which includes to begin moving from a state of rest), i.e., to accelerate, or a flexible object to deform, or both. Force can also be described by intuitive concepts such as a push or a pull. A force has both magnitude and direction, making it a vector quantity. It is measured in the SI unit of newtons and represented by the symbol F.

The original form of Newton's second law states that the net force acting upon an object is equal to the rate at which its momentum changes with time. If the mass of the object is constant, this law implies that the acceleration of an object is directly proportional to the net force acting on the object, is in the direction of the net force, and is inversely proportional to the mass of the object. As a formula, this is expressed as:

$$\vec{F} = m\vec{a}$$

where the arrows imply a vector quantity possessing both magnitude and direction.

Related concepts to force include: thrust, which increases the velocity of an object; drag, which decreases the velocity of an object; and torque which produces changes in rotational speed of an object. In an extended body, each part usually applies forces on the adjacent parts; the distribution of such forces through the body is the so-called mechanical stress. Pressure is a simple type of stress. Stress usually causes deformation of solid materials, or flow in fluids.

Development of the Concept

Philosophers in antiquity used the concept of force in the study of stationary and moving objects and simple machines, but thinkers such as Aristotle and Archimedes retained fundamental errors in understanding force. In part this was due to an incomplete understanding of the sometimes non-obvious force of friction, and a consequently inadequate view of the nature of natural motion. A fundamental error was the belief that a force is required to maintain motion, even at a constant velocity. Most of the previous misunderstandings about motion and force were eventually corrected by Sir Isaac Newton; with his mathematical insight, he formulated laws of motion that were not improved-on for nearly three hundred years. By the early 20th century, Einstein developed a theory of relativity that correctly predicted the action of forces on objects with increasing momenta near the speed of light, and also provided insight into the forces produced by gravitation and inertia.

With modern insights into quantum mechanics and technology that can accelerate particles close to the speed of light, particle physics has devised a Standard Model to describe forces between particles smaller than atoms. The Standard Model predicts that exchanged particles called gauge bosons are the fundamental means by which forces are emitted and absorbed. Only four main interactions are known: in order of decreasing strength, they are: strong, electromagnetic, weak, and gravitational. High-energy particle physics observations made during the 1970s and 1980s confirmed that the weak and electromagnetic forces are expressions of a more fundamental electroweak interaction.

Pre-Newtonian Concepts

Since antiquity the concept of force has been recognized as integral to the functioning of each of the simple machines. The mechanical advantage given by a simple machine allowed for less force to be used in exchange for that force acting over a greater distance for the same amount of work. Analysis of the characteristics of forces ultimately culminated in the work of Archimedes who was especially famous for formulating a treatment of buoyant forces inherent in fluids.

Aristotle provided a philosophical discussion of the concept of a force as an integral part of Aristotelian cosmology. In Aristotle's view, the natural world held four elements that existed in "natural states". Aristotle believed that it was the natural state of objects with mass on Earth, such as the elements water and earth, to be motionless on the ground and that they tended towards that state if left alone. He distinguished between the innate tendency of objects to find their "natural place" (e.g., for heavy bodies to fall), which led to "natural motion", and unnatural or forced motion, which required continued application of a force. This theory, based on the everyday experience of how objects move, such as the constant application of a force needed to keep a cart moving, had conceptual trouble accounting for the behaviour of projectiles, such as the flight of arrows. The place where forces were applied to projectiles was only at the start of the flight, and while the projectile sailed through the air, no discernible force acts on it. Aristotle was aware of this problem and proposed that the air displaced through the projectile's path provided the needed force to continue the projectile moving. This explanation demands that air is needed for projectiles and that, for example, in a vacuum, no projectile would move after the initial push. Additional problems with the explanation include the fact that air resists the motion of the projectiles.

Aristotelian physics began facing criticism in Medieval science, first by John Philoponus in the 6th century. The shortcomings of Aristotelian physics would not be fully corrected until the 17th century work of Galileo Galilei, who was influenced by the late Medieval idea that objects in forced motion carried an innate force of impetus. Galileo constructed an experiment in which stones and cannonballs were both rolled down an incline to disprove the Aristotelian theory of motion early in the 17th century. He showed that the bodies were accelerated by gravity to an extent which was independent of their mass and argued that objects retain their velocity unless acted on by a force, for example friction.

Newton's Laws of Motion

Newton's laws of motion are three physical laws that together laid the foundation for classical mechanics. They describe the relationship between a body and the forces acting upon it, and its motion in response to said forces. They have been expressed in several different ways over nearly three centuries, and can be summarized as follows:

1. *First law:* When viewed in an inertial reference frame, an object either is at rest or moves at a constant velocity, unless acted upon by a force.
2. *Second law:* The acceleration of a body is directly proportional to, and in the same direction as, the net force acting on the body, and inversely proportional to its mass. Thus, $F = ma$, where F is the net force acting on the object, m is the mass of the object and a is the acceleration of the object.
3. *Third law:* When one body exerts a force on a second body, the second body simultaneously exerts a force equal in magnitude and opposite in direction to that of the first body.

The three laws of motion were first compiled by Isaac Newton in his *Philosophiæ Naturalis Principia Mathematica* (*Mathematical Principles of Natural Philosophy*), first published in 1687. Newton used them to explain and investigate the motion of many physical objects and systems. For example, in the third volume of the text, Newton showed that these laws of motion, combined with his law of universal gravitation, explained Kepler's laws of planetary motion.

Overview

Newton's laws are applied to objects which are idealized as single point masses, in the sense that the size and shape of the object's body are neglected in order to focus on its motion more easily. This can be done when the object is small compared to the distances involved in its analysis, or the deformation and rotation of the body are of no importance. In this way, even a planet can be idealized as a particle for analysis of its orbital motion around a star.

In their original form, Newton's laws of motion are not adequate to characterize the motion of rigid bodies and deformable bodies. Leonard Euler in 1750 introduced a generalization of Newton's laws of motion for rigid bodies called the Euler's laws of motion, later applied as well for deformable bodies assumed as a continuum. If a body is represented as an assemblage of discrete particles, each governed by Newton's laws of motion, then Euler's laws can be derived from Newton's laws. Euler's laws can, however, be taken as axioms describing the laws of motion for extended bodies, independently of any particle structure.

Newton's laws hold only with respect to a certain set of frames of reference called Newtonian or inertial reference frames. Some authors interpret the first law as defining what an inertial reference frame is; from this point of view, the second law only holds when the observation is made from an inertial reference frame, and therefore the first law cannot be proved as a special case of the second. Other authors do treat the first law as a corollary of the second. The explicit concept of an inertial frame of reference was not developed until long after Newton's death.

In the given interpretation mass, acceleration, momentum, and (most importantly) force are assumed to be externally defined quantities. This is the most common, but not the only interpretation of the way one can consider the laws to be a definition of these quantities.

Newtonian mechanics has been superseded by special relativity, but it is still useful as an approximation when the speeds involved are much slower than the speed of light.

Newton's First Law

The first law states that if the net force (the vector sum of all forces acting on an object) is zero, then the velocity of the object is constant. Velocity is a vector quantity which expresses both the object's speed and the direction of its motion; therefore, the statement that the object's velocity is constant is a statement that both its speed and the direction of its motion are constant.

The first law can be stated mathematically as

$$\sum F = 0 \Rightarrow \frac{dv}{dt} = 0.$$

Consequently,

- An object that is at rest will stay at rest unless an external force acts upon it.
- An object that is in motion will not change its velocity unless an external force acts upon it.

This is known as *uniform motion*. An object *continues* to do whatever it happens to be doing unless a force is exerted upon it. If it is at rest, it continues in a state of rest (demonstrated when a tablecloth is skillfully whipped from under dishes on a tabletop and the dishes remain in their initial state of rest). If an object is moving, it continues to move without turning or changing its speed. This is evident in space probes that continually move in outer space. Changes in motion must be imposed against the tendency of an object to retain its state of motion. In the absence of net forces, a moving object tends to move along a straight line path indefinitely.

Newton placed the first law of motion to establish frames of reference for which the other laws are applicable. The first law of motion postulates the existence of at least one frame of reference called a Newtonian or inertial reference frame, relative to which the motion of a particle not subject to forces is a straight line at a constant speed. Newton's first law is often referred to as the *law of inertia*. Thus, a condition necessary for the uniform motion of a particle relative to an inertial reference frame is that the total net force acting on it is zero. In this sense, the first law can be restated as:

In every material universe, the motion of a particle in a preferential reference frame Φ is determined by the action of forces whose total vanished for all times when and only when the velocity of the particle is constant in Φ. That is, a particle initially at rest or in uniform motion in the preferential frame Φ continues in that state unless compelled by forces to change it.

Newton's laws are valid only in an inertial reference frame. Any reference frame that is in uniform motion with respect to an inertial frame is also an inertial frame, i.e. Galilean invariance or the principle of Newtonian relativity.

Newton's Second Law

The second law states that the net force on an object is equal to the rate of change (that is, the *derivative*) of its linear momentum p in an inertial reference frame:

$$F = \frac{dp}{dt} = \frac{d(mv)}{dt}.$$

The second law can also be stated in terms of an object's acceleration. Since the law is valid only for constant-mass systems, the mass can be taken outside the differentiation operator by the constant factor rule in differentiation. Thus,

$$F = m\frac{dv}{dt} = ma,$$

where *F* is the net force applied, *m* is the mass of the body, and a is the body's acceleration. Thus, the net force applied to a body produces a proportional acceleration. In other words, if a body is accelerating, then there is a force on it.

Consistent with the first law, the time derivative of the momentum is non-zero when the momentum changes direction, even if there is no change in its magnitude; such is the case with uniform circular motion. The relationship also implies the conservation of momentum: when the net force on the body is zero, the momentum of the body is constant. Any net force is equal to the rate of change of the momentum.

Any mass that is gained or lost by the system will cause a change in momentum that is not the result of an external force. A different equation is necessary for variable-mass systems.

Newton's second law requires modification if the effects of special relativity are to be taken into account, because at high speeds the approximation that momentum is the product of rest mass and velocity is not accurate.

Impulse

An impulse J occurs when a force *F* acts over an interval of time Δt, and it is given by

$$J = \int_{\Delta t} F\,dt.$$

Since force is the time derivative of momentum, it follows that

$$J = \Delta p = m\Delta v.$$

This relation between impulse and momentum is closer to Newton's wording of the second law.

Impulse is a concept frequently used in the analysis of collisions and impacts.

Variable-mass Systems

Variable-mass systems, like a rocket burning fuel and ejecting spent gases, are not closed and cannot be directly treated by making mass a function of time in the second law; that is, the following formula is wrong:

$$F_{net} = \frac{d}{dt}\left[m(t)v(t)\right] = m(t)\frac{dv}{dt} + v(t)\frac{dm}{dt}. \qquad \text{(wrong)}$$

The falsehood of this formula can be seen by noting that it does not respect Galilean invariance: a variable-mass object with $F = 0$ in one frame will be seen to have $F \neq 0$ in another frame.

The correct equation of motion for a body whose mass m varies with time by either ejecting or accreting mass is obtained by applying the second law to the entire, constant-mass system consisting of the body and its ejected/accreted mass; the result is

$$F + u\frac{dm}{dt} = m\frac{dv}{dt}$$

where u is the relative velocity of the escaping or incoming mass *as seen by the body*. From this equation one can derive the Tsiolkovsky rocket equation.

Under some conventions, the quantity $u\, dm/dt$ on the left-hand side, known as the thrust, is defined as a force (the force exerted on the body by the changing mass, such as rocket exhaust) and is included in the quantity F. Then, by substituting the definition of acceleration, the equation becomes $F = ma$.

Newton's Third Law

The third law states that all forces exist in pairs: if one object A exerts a force F_A on a second object B, then B simultaneously exerts a force F_B on A, and the two forces are equal and opposite: $F_A = -F_B$. The third law means that all forces are *interactions* between different bodies, and thus that there is no such thing as a unidirectional force or a force that acts on only one body. This law is sometimes referred to as the *action-reaction law*, with F_A called the "action" and F_B the "reaction". The action and the reaction are simultaneous, and it does not matter which is called the *action* and which is called *reaction*; both forces are part of a single interaction, and neither force exists without the other.

The two forces in Newton's third law are of the same type (e.g., if the road exerts a forward frictional force on an accelerating car's tires, then it is also a frictional force that Newton's third law predicts for the tires pushing backward on the road).

From a conceptual standpoint, Newton's third law is seen when a person walks: they push against the floor, and the floor pushes against the person. Similarly, the tires of a car push against the road while the road pushes back on the tires—the tires and road simultaneously push against each other. In swimming, a person interacts with the water, pushing the water backward, while the water simultaneously pushes the person forward—both the person and the water push against each other. The reaction forces account for the motion in these examples. These forces depend on friction; a person or car on ice, for example, may be unable to exert the action force to produce the needed reaction force.

History

Newton's 1st Law:

From the original Latin of Newton's *Principia*:

> "***Lex I:*** *Corpus omne perseverare in statu suo quiescendi vel movendi uniformiter in directum, nisi quatenus a viribus impressis cogitur statum illum mutare.*"

Translated to English, this reads:

> "***Law I:*** *Every body persists in its state of being at rest or of moving uniformly straight forward, except insofar as it is compelled to change its state by force impressed.*"

The ancient Greek philosopher Aristotle had the view that all objects have a natural place in the universe: that heavy objects (such as rocks) wanted to be at rest on the Earth and that light objects like smoke wanted to be at rest in the sky and the stars wanted to remain in the heavens. He thought that a body was in its natural state when it was at rest, and for the body to move in a straight line at a constant speed an external agent was needed to continually propel it, otherwise it would stop moving. Galileo Galilei, however, realized that a force is necessary to change the velocity of a body, i.e., acceleration, but no force is needed to maintain its velocity. In other words, Galileo stated that, in the *absence* of a force, a moving object will continue moving. The tendency of objects to resist changes in motion was what Galileo called *inertia*. This insight was refined by Newton, who made it into his first law, also known as the "law of inertia"—no force means no acceleration, and hence the body will maintain its velocity. As Newton's first law is a restatement of the law of inertia which Galileo had already described, Newton appropriately gave credit to Galileo.

The law of inertia apparently occurred to several different natural philosophers and scientists independently, including Thomas Hobbes in his *Leviathan*. The 17th century philosopher and mathematician René Descartes also formulated the law, although he did not perform any experiments to confirm it.

Newton's 2nd Law

Newton's original Latin reads:

> *"Lex II: Mutationem motus proportionalem esse vi motrici impressae, et fieri secundum lineam rectam qua vis illa imprimitur."*

This was translated quite closely in Motte's 1729 translation as:

> *"**Law II:** The alteration of motion is ever proportional to the motive force impress'd; and is made in the direction of the right line in which that force is impress'd."*

According to modern ideas of how Newton was using his terminology, this is understood, in modern terms, as an equivalent of:

> *The change of momentum of a body is proportional to the impulse impressed on the body, and happens along the straight line on which that impulse is impressed.*

Motte's 1729 translation of Newton's Latin continued with Newton's commentary on the second law of motion, reading:

> *If a force generates a motion, a double force will generate double the motion, a triple force triple the motion, whether that force be impressed altogether and at once, or gradually and successively. And this motion (being always directed the same way with the generating force), if the body moved before, is added to or subtracted from the former motion, according as they directly conspire with or are directly contrary to each other; or obliquely joined, when they are oblique, so as to produce a new motion compounded from the determination of both.*

The sense or senses in which Newton used his terminology, and how he understood the second law and intended it to be understood, have been extensively discussed by historians of science, along with the relations between Newton's formulation and modern formulations.

Newton's 3rd Law

> "**Lex III:** Actioni contrariam semper et æqualem esse reactionem: sive corporum duorum actiones in se mutuo semper esse æquales et in partes contrarias dirigi."
>
> "**Law III:** *To every action there is always an equal and opposite reaction: or the forces of two bodies on each other are always equal and are directed in opposite directions.*"

A more direct translation than the one just given above is:

LAW III: To every action there is always opposed an equal reaction: or the mutual actions of two bodies upon each other are always equal, and directed to contrary parts. — Whatever draws or presses another is as much drawn or pressed by that other. If you press a stone with your finger, the finger is also pressed by the stone. If a horse draws a stone tied to a rope, the horse (if I may so say) will be equally drawn back towards the stone: for the distended rope, by the same endeavour to relax or unbend itself, will draw the horse as much towards the stone, as it does the stone towards the horse, and will obstruct the progress of the one as much as it advances that of the other. If a body impinges upon another, and by its force changes the motion of the other, that body also (because of the equality of the mutual pressure) will undergo an equal change, in its own motion, toward the contrary part. The changes made by these actions are equal, not in the velocities but in the motions of the bodies; that is to say, if the bodies are not hindered by any other impediments. For, as the motions are equally changed, the changes of the velocities made toward contrary parts are reciprocally proportional to the bodies. This law takes place also in attractions, as will be proved in the next scholium.

In the above, as usual, *motion* is Newton's name for momentum, hence his careful distinction between motion and velocity.

Newton used the third law to derive the law of conservation of momentum; however from a deeper perspective, conservation of momentum is the more fundamental idea (derived via Noether's theorem from Galilean invariance), and holds in cases where Newton's third law appears to fail, for instance when force fields as well as particles carry momentum, and in quantum mechanics.

Importance and Range of Validity

Newton's laws were verified by experiment and observation for over 200 years, and they are excellent approximations at the scales and speeds of everyday life. Newton's laws of motion, together with his law of universal gravitation and the mathematical techniques of calculus, provided for the first time a unified quantitative explanation for a wide range of physical phenomena.

These three laws hold to a good approximation for macroscopic objects under everyday conditions. However, Newton's laws (combined with universal gravitation and classical electrodynamics) are inappropriate for use in certain circumstances, most notably at very small scales, very high speeds (in special relativity, the Lorentz factor must be included in the expression for momentum along with rest mass and velocity) or very strong gravitational fields. Therefore, the laws cannot be used to explain phenomena such as conduction of electricity in a semiconductor, optical properties of substances, errors in non-relativistically corrected GPS systems and superconductivity. Explanation of these phenomena requires more sophisticated physical theories, including general relativity and quantum field theory.

In quantum mechanics concepts such as force, momentum, and position are defined by linear operators that operate on the quantum state; at speeds that are much lower than the speed of light, Newton's laws are just as exact for these operators as they are for classical objects. At speeds comparable to the speed of light, the second law holds in the original form $F = dp/dt$, where F and p are four-vectors.

Relationship to the Conservation Laws

In modern physics, the laws of conservation of momentum, energy, and angular momentum are of more general validity than Newton's laws, since they apply to both light and matter, and to both classical and non-classical physics.

This can be stated simply, "Momentum, energy and angular momentum cannot be created or destroyed."

Because force is the time derivative of momentum, the concept of force is redundant and subordinate to the conservation of momentum, and is not used in fundamental theories (e.g., quantum mechanics, quantum electrodynamics, general relativity, etc.). The standard model explains in detail how the three fundamental forces known as gauge forces originate out of exchange by virtual particles. Other forces such as gravity and fermionic degeneracy pressure also arise from the momentum conservation. Indeed, the conservation of 4-momentum in inertial motion via curved space-time results in what we call gravitational force in general relativity theory. Application of space derivative (which is a momentum operator in quantum mechanics) to overlapping wave functions of pair of fermions (particles with half-integer spin) results in shifts of maxima of compound wavefunction away from each other, which is observable as "repulsion" of fermions.

Newton stated the third law within a world-view that assumed instantaneous action at a distance between material particles. However, he was prepared for philosophical criticism of this action at a distance, and it was in this context that he stated the famous phrase "I feign no hypotheses". In modern physics, action at a distance has been completely eliminated, except for subtle effects involving quantum entanglement. However in modern engineering in all practical applications involving the motion of vehicles and satellites, the concept of action at a distance is used extensively.

The discovery of the Second Law of Thermodynamics by Carnot in the 19th century showed that every physical quantity is not conserved over time, thus disproving the validity of inducing the opposite metaphysical view from Newton's laws. Hence, a "steady-state" worldview based solely on Newton's laws and the conservation laws does not take entropy into account.

Descriptions

Since forces are perceived as pushes or pulls, this can provide an intuitive understanding for describing forces. As with other physical concepts (e.g. temperature), the intuitive understanding of forces is quantified using precise operational definitions that are consistent with direct observations and compared to a standard measurement scale. Through experimentation, it is determined that laboratory measurements of forces are fully consistent with the conceptual definition of force offered by Newtonian mechanics.

Forces act in a particular direction and have sizes dependent upon how strong the push or pull is. Because of these characteristics, forces are classified as "vector quantities". This means that forces follow a different set of mathematical rules than physical quantities that do not have direction (denoted scalar quantities). For example, when determining what happens when two forces act on the same object, it is necessary to know both the magnitude and the direction of both forces to calculate the result.

If both of these pieces of information are not known for each force, the situation is ambiguous. For example, if you know that two people are pulling on the same rope with known magnitudes of force but you do not know which direction either person is pulling, it is impossible to determine what the acceleration of the rope will be.

The two people could be pulling against each other as in tug of war or the two people could be pulling in the same direction. In this simple one-dimensional example, without knowing the direction of the forces it is impossible to decide whether the net force is the result of adding the two force magnitudes or subtracting one from the other. Associating forces with vectors avoids such problems.

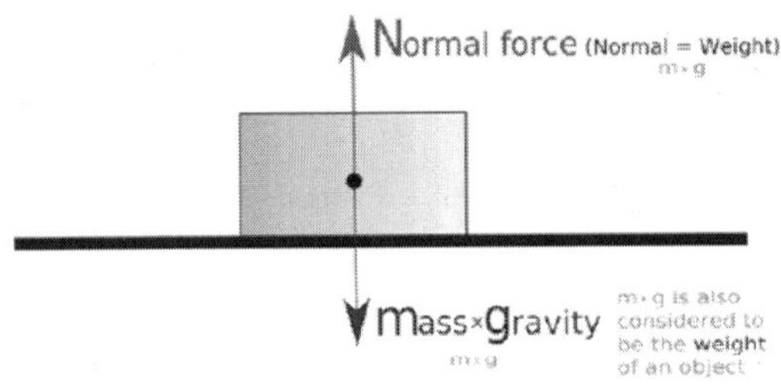

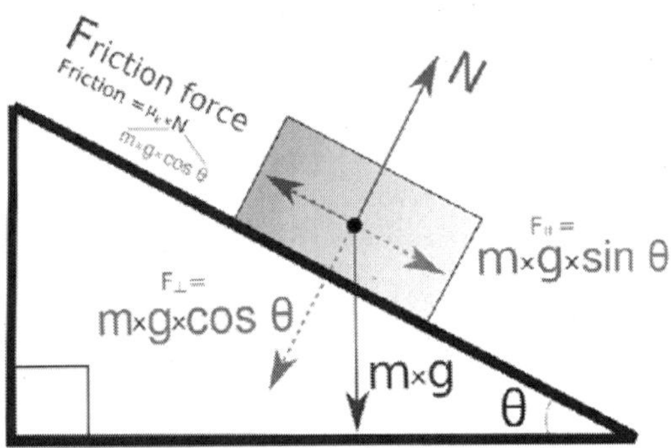

Figure: Diagrams of a block on a flat surface and an inclined plane. Forces are resolved and added together to determine their magnitudes and the net force.

Historically, forces were first quantitatively investigated in conditions of static equilibrium where several forces canceled each other out. Such experiments demonstrate the crucial properties that forces are additive vector quantities: they have magnitude and direction. When two forces act on a point particle, the resulting force, the *resultant* (also called the *net force*), can be determined by following the parallelogram rule of vector addition: the addition of two vectors represented by sides of a parallelogram, gives an equivalent resultant vector which is equal in magnitude and direction to the transversal of the parallelogram.

The magnitude of the resultant varies from the difference of the magnitudes of the two forces to their sum, depending on the angle between their lines of action. However, if the forces are acting on an extended body, their respective lines of application must also be specified in order to account for their effects on the motion of the body.

Free-body diagrams can be used as a convenient way to keep track of forces acting on a system. Ideally, these diagrams are drawn with the angles and relative magnitudes of the force vectors preserved so that graphical vector addition can be done to determine the net force.

As well as being added, forces can also be resolved into independent components at right angles to each other. A horizontal force pointing northeast can therefore be split into two forces, one pointing north, and one pointing east. Summing these component forces using vector addition yields the original force.

Resolving force vectors into components of a set of basis vectors is often a more mathematically clean way to describe forces than using magnitudes and directions. This is because, for orthogonal components, the components of the vector sum are uniquely determined by the scalar addition of the components of the individual vectors.

Orthogonal components are independent of each other because forces acting at ninety degrees to each other have no effect on the magnitude or direction of the other. Choosing a set of orthogonal basis vectors is often done by considering what set of basis vectors will make the mathematics most convenient. Choosing a basis vector that is in the same direction as one of the forces is desirable, since that force would then have only one non-zero component. Orthogonal force vectors can be three-dimensional with the third component being at right-angles to the other two.

Equilibrium

Equilibrium occurs when the resultant force acting on a point particle is zero (that is, the vector sum of all forces is zero). When dealing with an extended body, it is also necessary that the net torque in it is 0. There are two kinds of equilibrium: static equilibrium and dynamic equilibrium.

Static Equilibrium

Static equilibrium was understood well before the invention of classical mechanics. Objects which are at rest have zero net force acting on them.

The simplest case of static equilibrium occurs when two forces are equal in magnitude but opposite in direction. For example, an object on a level surface is pulled (attracted) downward toward the centre of the Earth by the force of gravity. At the same time, surface forces resist the downward force with equal upward force (called the normal force). The situation is one of zero net force and no acceleration.

Pushing against an object on a frictional surface can result in a situation where the object does not move because the applied force is opposed by static friction, generated between the object and the table surface. For a situation with no movement, the static friction force *exactly* balances the applied force resulting in no acceleration. The static friction increases or decreases in response to the applied force up to an upper limit determined by the characteristics of the contact between the surface and the object.

A static equilibrium between two forces is the most usual way of measuring forces, using simple devices such as weighing scales and spring balances. For example, an object suspended on a vertical spring scale experiences the force of gravity acting on the object balanced by a force applied by the "spring reaction force" which equals the object's weight. Using such tools, some quantitative force laws were discovered: that the force of gravity is proportional to volume for objects of constant density (widely exploited for millennia to define standard weights); Archimedes' principle for buoyancy; Archimedes' analysis of the lever; Boyle's law for gas pressure; and Hooke's law for springs. These were all formulated and experimentally verified before Isaac Newton expounded his Three Laws of Motion.

Dynamic Equilibrium

Dynamic equilibrium was first described by Galileo who noticed that certain assumptions of Aristotelian physics were contradicted by observations and logic. Galileo realized that simple velocity addition demands that the concept of an "absolute rest frame" did not exist. Galileo concluded that motion in a constant velocity was completely equivalent to rest. This was contrary to Aristotle's notion of a "natural state" of rest that objects with mass naturally approached. Simple experiments showed that Galileo's understanding of the equivalence of constant velocity and rest were correct. For example, if a mariner dropped a cannonball from the crow's nest of a ship moving at a constant velocity, Aristotelian physics would have the cannonball fall straight down while the ship moved beneath it.

Thus, in an Aristotelian universe, the falling cannonball would land behind the foot of the mast of a moving ship. However, when this experiment is actually conducted, the cannonball always falls at the foot of the mast, as if the cannonball knows to travel with the ship despite being separated from it. Since there is no forward horizontal force being applied on the cannonball as it falls, the only conclusion left is that the cannonball continues to move with the same velocity as the boat as it falls. Thus, no force is required to keep the cannonball moving at the constant forward velocity.

Moreover, any object travelling at a constant velocity must be subject to zero net force (resultant force). This is the definition of dynamic equilibrium: when all the forces on an object balance but it still moves at a constant velocity.

A simple case of dynamic equilibrium occurs in constant velocity motion across a surface with kinetic friction. In such a situation, a force is applied in the direction of motion while the kinetic friction force exactly opposes the applied force. This results in zero net force, but since the object started with a non-zero velocity, it continues to move with a non-zero velocity. Aristotle misinterpreted this motion as being caused by the applied force. However, when kinetic friction is taken into consideration it is clear that there is no net force causing constant velocity motion.

Special Relativity

In the special theory of relativity, mass and energy are equivalent (as can be seen by calculating the work required to accelerate an object). When an object's velocity increases, so does its energy and hence its mass equivalent (inertia). It thus requires more force to accelerate it the same amount than it did at a lower velocity. Newton's Second Law

$$\vec{F} = \mathrm{d}\vec{p} / \mathrm{d}t$$

remains valid because it is a mathematical definition. But in order to be conserved, relativistic momentum must be redefined as:

$$\vec{p} = \frac{m_0 \vec{v}}{\sqrt{1 - v^2 / c^2}}$$

where

$v$is the velocity and

$c$is the speed of light

m_0 is the rest mass.

The relativistic expression relating force and acceleration for a particle with constant non-zero rest mass m moving in the x direction is:

$$F_x = \gamma^3 m a_x$$

$$F_y = \gamma m a_y$$

$$F_z = \gamma m a_z$$

where the Lorentz factor

$$\gamma = \frac{1}{\sqrt{1 - v^2 / c^2}}.$$

In the early history of relativity, the expressions $\gamma^3 m$ and γm were called longitudinal and transverse mass. Relativistic force does not produce a constant acceleration, but an ever decreasing acceleration as the object approaches the speed of light. Note that γ is undefined for an object with a non-zero rest mass at the speed of light, and the theory yields no prediction at that speed.

One can, however, restore the form of

$$F^{\mu} = mA^{\mu}$$

for use in relativity through the use of four-vectors. This relation is correct in relativity when F^{μ} is the four-force, m is the invariant mass, and A^{μ} is the four-acceleration.

Feynman Diagrams

In modern particle physics, forces and the acceleration of particles are explained as a mathematical by-product of exchange of momentum-carrying gauge bosons. With the development of quantum field theory and general relativity, it was realized that force is a redundant concept arising from conservation of momentum (4-momentum in relativity and momentum of virtual particles in quantum electrodynamics). The conservation of momentum, can be directly derived from homogeneity (=shift symmetry) of space and so is usually considered more fundamental than the concept of a force.

Thus the currently known fundamental forces are considered more accurately to be "fundamental interactions". When particle A emits (creates) or absorbs (annihilates) virtual particle B, a momentum conservation results in recoil of particle A making impression of repulsion or attraction between particles A A' exchanging by B. This description applies to all forces arising from fundamental interactions. While sophisticated mathematical descriptions are needed to predict, in full detail, the accurate result of such interactions, there is a conceptually simple way to describe such interactions through the use of Feynman diagrams. In a Feynman diagram, each matter particle is represented as a straight line travelling through time which normally increases up or to the right in the diagram. Matter and anti-matter particles are identical except for their direction of propagation through the Feynman diagram. World lines of particles intersect at interaction vertices, and the Feynman diagram represents any force arising from an interaction as occurring at the vertex with an associated instantaneous change in the direction of the particle world lines. Gauge bosons are emitted away from the vertex as wavy lines and, in the case of virtual particle exchange, are absorbed at an adjacent vertex.

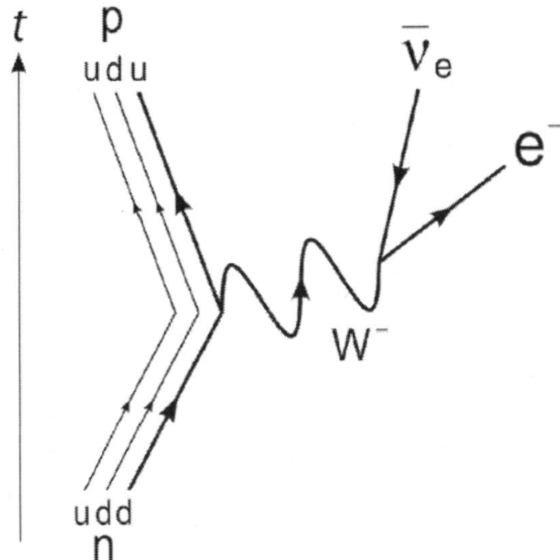

Figure: *Feynman diagram for the decay of a neutron into a proton. The W boson is between two vertices indicating a repulsion.*

The utility of Feynman diagrams is that other types of physical phenomena that are part of the general picture of fundamental interactions but are conceptually separate from forces can also be described using the same rules. For example, a Feynman diagram can describe in succinct detail how a neutron decays into an electron, proton, and neutrino, an interaction mediated by the same gauge boson that is responsible for the weak nuclear force.

Fundamental Models

All the forces in the universe are based on four fundamental interactions. The strong and weak forces act only at very short distances, and are responsible for the interactions between subatomic particles including nucleons and compound nuclei. The electromagnetic force acts between electric charges and the gravitational force acts between masses. All other forces are based on the existence of the four fundamental interactions. For example, friction is a manifestation of the electromagnetic force acting between the atoms of two surfaces, and the Pauli Exclusion Principle, which does not allow atoms to pass through each other. The forces in springs, modelled by Hooke's law, are also the result of electromagnetic forces and the Exclusion Principle acting together to return the object to its equilibrium position. Centrifugal forces are acceleration forces which arise simply from the acceleration of rotating frames of reference.

The development of fundamental theories for forces proceeded along the lines of unification of disparate ideas. For example, Isaac Newton unified the force responsible for objects falling at the surface of the Earth with the force responsible for the orbits of celestial mechanics in his universal theory of gravitation. Michael Faraday and James Clerk Maxwell demonstrated that electric and magnetic forces were unified through one consistent theory of electromagnetism. In the 20th century, the development of quantum mechanics led to a modern understanding that the first three fundamental forces (all except gravity) are manifestations of matter (fermions) interacting by exchanging virtual particles called gauge bosons. This standard model of particle physics posits a similarity between the forces and led scientists to predict the unification of the weak and electromagnetic forces in electroweak theory subsequently confirmed by observation. The complete formulation of the standard model predicts an as yet unobserved Higgs mechanism, but observations such as neutrino oscillations indicate that the standard model is incomplete. A grand unified theory allowing for the combination of the electroweak interaction with the strong force is held out as a possibility with candidate theories such as supersymmetry proposed to accommodate some of the outstanding unsolved problems in physics. Physicists are still attempting to develop self-consistent unification models that would combine all four fundamental interactions into a theory of everything. Einstein tried and failed at this endeavour, but currently the most popular approach to answering this question is string theory.

Gravity

What we now call gravity was not identified as a universal force until the work of Isaac Newton. Before Newton, the tendency for objects to fall towards the Earth was not understood to be related to the motions of celestial objects. Galileo was instrumental in describing the characteristics of falling objects by determining that the acceleration of every object in free-fall was constant and independent of the mass of the object. Today, this acceleration due to gravity towards the surface of the Earth is usually designated as $\vec{g}$ and has a magnitude of about 9.81 meters per second squared (this measurement is taken from sea level and may vary depending on location), and points toward the centre of the Earth. This observation means that the force of gravity on an object at the Earth's surface is directly proportional to the object's mass. Thus an object that has a mass of m will experience a force:

$$\vec{F} = m\vec{g}$$

In free-fall, this force is unopposed and therefore the net force on the object is its weight. For objects not in free-fall, the force of gravity is opposed by the reactions of their supports. For example, a person standing on the ground experiences zero net force, since his weight is balanced by a normal force exerted by the ground.

Newton's contribution to gravitational theory was to unify the motions of heavenly bodies, which Aristotle had assumed were in a natural state of constant motion, with falling motion observed on the Earth. He proposed a law of gravity that could account for the celestial motions that had been described earlier using Kepler's Laws of Planetary Motion.

Newton came to realize that the effects of gravity might be observed in different ways at larger distances. In particular, Newton determined that the acceleration of the Moon around the Earth could be ascribed to the same force of gravity if the acceleration due to gravity decreased as an inverse square law. Further, Newton realized that the acceleration due to gravity is proportional to the mass of the attracting body. Combining these ideas gives a formula that relates the mass ($m_{\oplus}$) and the radius ($R_{\oplus}$) of the Earth to the gravitational acceleration:

$$\vec{g} = -\frac{Gm_{\oplus}}{R_{\oplus}^{2}}\hat{r}$$

where the vector direction is given by $\hat{r}$, the unit vector directed outward from the centre of the Earth.

In this equation, a dimensional constant G is used to describe the relative strength of gravity. This constant has come to be known as Newton's Universal Gravitation Constant, though its value was unknown in Newton's lifetime. Not until 1798 was Henry Cavendish able to make the first measurement of G using a torsion balance; this was widely reported in the press as a measurement of the mass of the Earth since knowing G could allow one to solve for the Earth's mass given the above equation. Newton, however, realized that since all celestial bodies followed the same laws of motion, his law of gravity had to be universal. Succinctly stated, Newton's Law of Gravitation states that the force on a spherical object of mass m_1 due to the gravitational pull of mass m_2 is

$$\vec{F} = -\frac{Gm_1m_2}{r^2}\hat{r}$$

where r is the distance between the two objects' centres of mass and $\hat{r}$ is the unit vector pointed in the direction away from the centre of the first object toward the centre of the second object.

This formula was powerful enough to stand as the basis for all subsequent descriptions of motion within the solar system until the 20th century. During that time, sophisticated methods of perturbation analysis were invented to calculate the deviations of orbits due to the influence of multiple bodies on a planet, moon, comet, or asteroid. The formalism was exact enough to allow mathematicians to predict the existence of the planet Neptune before it was observed.

It was only the orbit of the planet Mercury that Newton's Law of Gravitation seemed not to fully explain. Some astrophysicists predicted the existence of another planet (Vulcan) that would explain the discrepancies; however, despite some early indications, no such planet could be found. When Albert Einstein finally formulated his theory of general relativity (GR) he turned his attention to the problem of Mercury's orbit and found that his theory added a correction which could account for the discrepancy. This was the first time that Newton's Theory of Gravity had been shown to be less correct than an alternative.

Since then, and so far, general relativity has been acknowledged as the theory which best explains gravity. In GR, gravitation is not viewed as a force, but rather, objects moving freely in gravitational fields travel under their own inertia in straight lines through curved space-time – defined as the shortest space-time path between two space-time events. From the perspective of the object, all motion occurs as if there were no gravitation whatsoever. It is only when observing the motion in a global sense that the curvature of space-time can be observed and the force is inferred from the object's curved path. Thus, the straight line path in space-time is seen as a curved line in space, and it is called the *ballistic trajectory* of the object. For example, a basketball thrown from the ground moves in a parabola, as it is in a uniform gravitational field. Its space-time trajectory (when the extra ct dimension is added) is almost a straight line, slightly curved (with the radius of curvature of the order of few light-years). The time derivative of the changing momentum of the object is what we label as "gravitational force".

Electromagnetic Forces

The electrostatic force was first described in 1784 by Coulomb as a force which existed intrinsically between two charges. The properties of the electrostatic force were that it varied as an inverse square law directed in the radial direction, was both attractive and repulsive (there was intrinsic polarity), was independent of the mass of the charged objects, and followed the superposition principle. Coulomb's Law unifies all these observations into one succinct statement.

Subsequent mathematicians and physicists found the construct of the *electric field* to be useful for determining the electrostatic force on an electric charge at any point in space. The electric field was based on using a hypothetical "test charge" anywhere in space and then using Coulomb's Law to determine the electrostatic force. Thus the electric field anywhere in space is defined as

$$\vec{E} = \frac{\vec{F}}{q}$$

where q is the magnitude of the hypothetical test charge.

Meanwhile, the Lorentz force of magnetism was discovered to exist between two electric currents. It has the same mathematical character as Coulomb's Law with the proviso that like currents attract and unlike currents repel. Similar to the electric field, the magnetic field can be used to determine the magnetic force on an electric current at any point in space. In this case, the magnitude of the magnetic field was determined to be

$$B = \frac{F}{I\ell}$$

where I is the magnitude of the hypothetical test current and ℓ is the length of hypothetical wire through which the test current flows. The magnetic field exerts a force on all magnets including, for example, those used in compasses. The fact that the Earth's magnetic field is aligned closely with the orientation of the Earth's axis causes compass magnets to become oriented because of the magnetic force pulling on the needle.

Through combining the definition of electric current as the time rate of change of electric charge, a rule of vector multiplication called Lorentz's Law describes the force on a charge moving in a magnetic field. The connection between electricity and magnetism allows for the description of a unified *electromagnetic force* that acts on a charge. This force can be written as a sum of the electrostatic force (due to the electric field) and the magnetic force (due to the magnetic field). Fully stated, this is the law:

$$\vec{F} = q(\vec{E} + \vec{v} \times \vec{B})$$

where $\vec{F}$ is the electromagnetic force, q is the magnitude of the charge of the particle, $\vec{E}$ is the electric field, $\vec{v}$ is the velocity of the particle which is crossed with the magnetic field ($\vec{B}$).

The origin of electric and magnetic fields would not be fully explained until 1864 when James Clerk Maxwell unified a number of earlier theories into a set of 20 scalar equations, which were later reformulated into 4 vector equations by Oliver Heaviside and Josiah Willard Gibbs. These "Maxwell Equations" fully described the sources of the fields as being stationary and moving charges, and the interactions of the fields themselves. This led Maxwell to discover that electric and magnetic fields could be "self-generating" through a wave that travelled at a speed which he calculated to be the speed of light. This insight united the nascent fields of electromagnetic theory with optics and led directly to a complete description of the electromagnetic spectrum.

However, attempting to reconcile electromagnetic theory with two observations, the photoelectric effect, and the nonexistence of the ultraviolet catastrophe, proved troublesome. Through the work of leading theoretical physicists, a new theory of electromagnetism was developed using quantum mechanics. This final modification to electromagnetic theory ultimately led to quantum electrodynamics (or QED), which fully describes all electromagnetic phenomena as being mediated by wave-particles known as photons. In QED, photons are the fundamental exchange particle which described all interactions relating to electromagnetism including the electromagnetic force.

It is a common misconception to ascribe the stiffness and rigidity of solid matter to the repulsion of like charges under the influence of the electromagnetic force. However, these characteristics actually result from the Pauli Exclusion Principle. Since electrons are fermions, they cannot occupy the same quantum mechanical state as other electrons. When the electrons in a material are densely packed together, there are not enough lower energy quantum mechanical states for them all, so some of them must be in higher energy states. This means that it takes energy to pack them together. While this effect is manifested macroscopically as a structural force, it is technically only the result of the existence of a finite set of electron states.

Nuclear Forces

There are two "nuclear forces" which today are usually described as interactions that take place in quantum theories of particle physics. The strong nuclear force is the force responsible for the structural integrity of atomic nuclei while the weak nuclear force is responsible for the decay of certain nucleons into leptons and other types of hadrons.

The strong force is today understood to represent the interactions between quarks and gluons as detailed by the theory of quantum chromodynamics (QCD). The strong force is the fundamental force mediated by gluons, acting upon quarks, antiquarks, and the gluons themselves. The (aptly named) strong interaction is the "strongest" of the four fundamental forces.

The strong force only acts *directly* upon elementary particles. However, a residual of the force is observed between hadrons (the best known example being the force that acts between nucleons in atomic nuclei) as the nuclear force. Here the strong force acts indirectly, transmitted as gluons which form part of the virtual pi and rho mesons which classically transmit the nuclear force. The failure of many searches for free quarks has shown that the elementary particles affected are not directly observable. This phenomenon is called colour confinement.

The weak force is due to the exchange of the heavy W and Z bosons. Its most familiar effect is beta decay (of neutrons in atomic nuclei) and the associated radioactivity. The word "weak" derives from the fact that the field strength is some 10^{13} times less than that of the strong force. Still, it is stronger than gravity over short distances. A consistent electroweak theory has also been developed which shows that electromagnetic forces and the weak force are indistinguishable at a temperatures in excess of approximately 10^{15} kelvins. Such temperatures have been probed in modern particle accelerators and show the conditions of the universe in the early moments of the Big Bang.

Non-fundamental Forces

Some forces are consequences of the fundamental ones. In such situations, idealized models can be utilized to gain physical insight.

Normal Force

The normal force is due to repulsive forces of interaction between atoms at close contact. When their electron clouds overlap, Pauli repulsion (due to fermionic nature of electrons) follows resulting in the force which acts in a direction normal to the surface interface between two objects. The normal force, for example, is responsible for the structural integrity of tables and floors as well as being the force that responds whenever an external force pushes on a solid object. An example of the normal force in action is the impact force on an object crashing into an immobile surface.

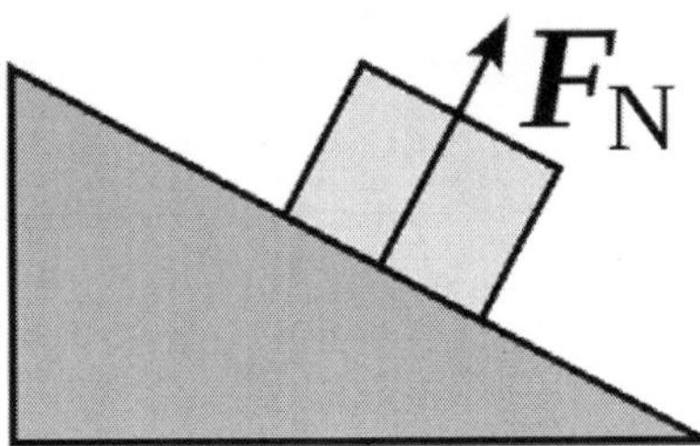

Figure: F_N *represents the normal force exerted on the object.*

Friction

Friction is a surface force that opposes relative motion. The frictional force is directly related to the normal force which acts to keep two solid objects separated at the point of contact. There are two broad classifications of frictional forces: static friction and kinetic friction. The static friction force (F_{sf}) will exactly oppose forces applied to an object parallel to a surface contact up to the limit specified by the coefficient of static friction (μ_{sf}) multiplied by the normal force (F_N). In other words the magnitude of the static friction force satisfies the inequality:

$$0 \leq F_{sf} \leq \mu_{sf} F_N .$$

The kinetic friction force (F_{kf}) is independent of both the forces applied and the movement of the object. Thus, the magnitude of the force equals:

$$F_{kf} = \mu_{kf} F_N ,$$

where μ_{kf} is the coefficient of kinetic friction. For most surface interfaces, the coefficient of kinetic friction is less than the coefficient of static friction.

Tension

Tension forces can be modelled using ideal strings which are massless, frictionless, unbreakable, and unstretchable. They can be combined with ideal pulleys which allow ideal strings to switch physical direction. Ideal strings transmit tension forces instantaneously in action-reaction pairs so that if two objects are connected by an ideal string, any force directed along the string by the first object is accompanied by a force directed along the string in the opposite direction by the second object. By connecting the same string multiple times to the same object through the use of a set-up that uses movable pulleys, the tension force on a load can be multiplied. For every string that acts on a load, another factor of the tension force in the string acts on the load. However, even though such machines allow for an increase in force, there is a corresponding increase in the length of string that must be displaced in order to move the load. These tandem effects result ultimately in the conservation of mechanical energy since the work done on the load is the same no matter how complicated the machine.

Elastic Force

An elastic force acts to return a spring to its natural length. An ideal spring is taken to be massless, frictionless, unbreakable, and infinitely stretchable. Such springs exert forces that push when contracted, or pull when extended, in proportion to the displacement of the spring from its equilibrium position.

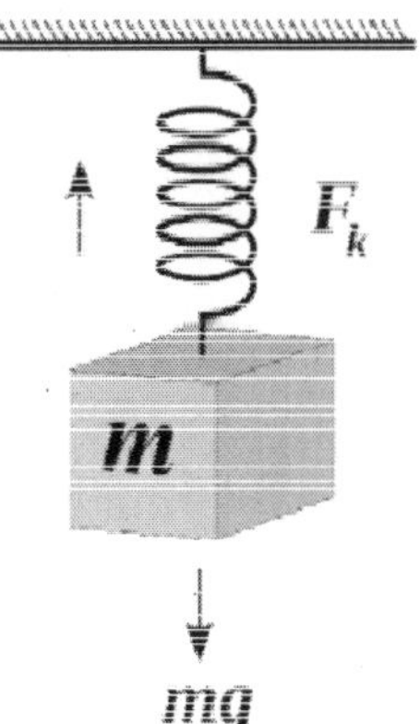

Figure: F_k *is the force that responds to the load on the spring*

This linear relationship was described by Robert Hooke in 1676, for whom Hooke's law is named. If Δx is the displacement, the force exerted by an ideal spring equals:

$$\vec{F} = -k\Delta\vec{x}$$

where k is the spring constant (or force constant), which is particular to the spring. The minus sign accounts for the tendency of the force to act in opposition to the applied load.

Continuum Mechanics

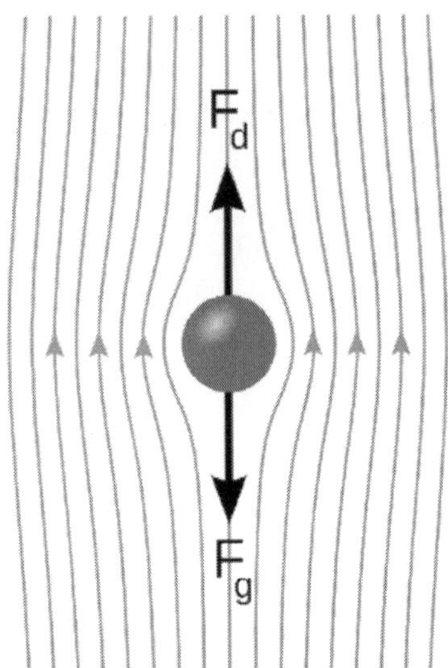

Figure: *When the drag force (F_d) associated with air resistance becomes equal in magnitude to the force of gravity on a falling object (F_g), the object reaches a state of dynamic equilibrium at terminal velocity.*

Newton's laws and Newtonian mechanics in general were first developed to describe how forces affect idealized point particles rather than three-dimensional objects. However, in real life, matter has extended structure and forces that act on one part of an object might affect other parts of an object. For situations where lattice holding together the atoms in an object is able to flow, contract, expand, or otherwise change shape, the theories of continuum mechanics describe the way forces affect the material. For example, in extended fluids, differences in pressure result in forces being directed along the pressure gradients as follows:

$$\frac{\vec{F}}{V} = -\vec{\nabla} P$$

where V is the volume of the object in the fluid and P is the scalar function that describes the pressure at all locations in space. Pressure gradients and differentials result in the buoyant force for fluids suspended in gravitational fields, winds in atmospheric science, and the lift associated with aerodynamics and flight.

A specific instance of such a force that is associated with dynamic pressure is fluid resistance: a body force that resists the motion of an object through a fluid due to viscosity. For so-called "Stokes' drag" the force is approximately proportional to the velocity, but opposite in direction:

$$\vec{F}_{\mathrm{d}} = -b\vec{v}$$

where:

b is a constant that depends on the properties of the fluid and the dimensions of the object (usually the cross-sectional area), and

$\vec{v}$ is the velocity of the object.

More formally, forces in continuum mechanics are fully described by a stress-tensor with terms that are roughly defined as

$$\sigma = \frac{F}{A}$$

where A is the relevant cross-sectional area for the volume for which the stress-tensor is being calculated. This formalism includes pressure terms associated with forces that act normal to the cross-sectional area (the matrix diagonals of the tensor) as well as shear terms associated with forces that act parallel to the cross-sectional area (the off-diagonal elements). The stress tensor accounts for forces that cause all deformations including also tensile stresses and compressions.

Fictitious forces

There are forces which are frame dependent, meaning that they appear due to the adoption of non-Newtonian (that is, non-inertial) reference frames. Such forces include the centrifugal force and the Coriolis force. These forces are considered fictitious because they do not exist in frames of reference that are not accelerating. In general relativity, gravity becomes a fictitious force that arises in situations where spacetime deviates from a flat geometry. As an extension, Kaluza-Klein theory and string theory ascribe electromagnetism and the other fundamental forces respectively to the curvature of differently scaled dimensions, which would ultimately imply that all forces are fictitious.

Rotations and Torque

Forces that cause extended objects to rotate are associated with torques. Mathematically, the torque of a force $\vec{F}$ is defined relative to an arbitrary reference point as the cross-product:

$$\vec{\tau} = \vec{r} \times \vec{F}$$

where

$\vec{r}$ is the position vector of the force application point relative to the reference point.

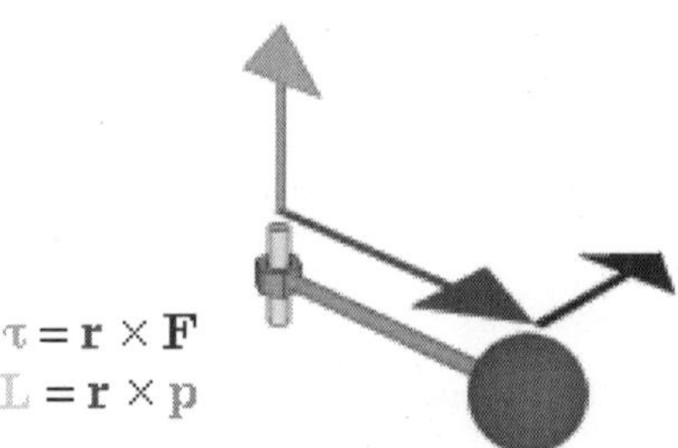

Figure: *Relationship between force (F), torque (τ), and momentum vectors (p and L) in a rotating system.*

Torque is the rotation equivalent of force in the same way that angle is the rotational equivalent for position, angular velocity for velocity, and angular momentum for momentum. As a consequence of Newton's First Law of Motion, there exists rotational inertia that ensures that all bodies maintain their angular momentum unless acted upon by an unbalanced torque. Likewise, Newton's Second Law of Motion can be used to derive an analogous equation for the instantaneous angular acceleration of the rigid body:

$$\vec{\tau} = I\vec{\alpha}$$

where

I is the moment of inertia of the body

$\vec{\alpha}$ is the angular acceleration of the body.

This provides a definition for the moment of inertia which is the rotational equivalent for mass. In more advanced treatments of mechanics, where the rotation over a time interval is described, the moment of inertia must be substituted by the tensor that, when properly analyzed, fully determines the characteristics of rotations including precession and nutation.

Equivalently, the differential form of Newton's Second Law provides an alternative definition of torque:

$$\vec{\tau} = \frac{d\vec{L}}{dt},$$

where $\vec{L}$ is the angular momentum of the particle.

Newton's Third Law of Motion requires that all objects exerting torques themselves experience equal and opposite torques, and therefore also directly implies the conservation of angular momentum for closed systems that experience rotations and revolutions through the action of internal torques.

Centripetal Force

For an object accelerating in circular motion, the unbalanced force acting on the object equals:

$$\vec{F} = -\frac{mv^2\hat{r}}{r}$$

where m is the mass of the object, $v$is the velocity of the object and r is the distance to the centre of the circular path and $\hat{r}$ is the unit vector pointing in the radial direction outwards from the centre. This means that the unbalanced centripetal force felt by any object is always directed toward the centre of the curving path. Such forces act perpendicular to the velocity vector associated with the motion of an object, and therefore do not change the speed of the object (magnitude of the velocity), but only the direction of the velocity vector. The unbalanced force that accelerates an object can be resolved into a component that is perpendicular to the path, and one that is tangential to the path. This yields both the tangential force which accelerates the object by either slowing it down or speeding it up and the radial (centripetal) force which changes its direction.

Kinematic Integrals

Forces can be used to define a number of physical concepts by integrating with respect to kinematic variables. For example, integrating with respect to time gives the definition of impulse:

$$\vec{I} = \int_{t_1}^{t_2} \vec{F}\mathrm{d}t$$

which, by Newton's Second Law, must be equivalent to the change in momentum (yielding the Impulse momentum theorem).

Similarly, integrating with respect to position gives a definition for the work done by a force:

$$W = \int_{\vec{x}_1}^{\vec{x}_2} \vec{F}\cdot\mathrm{d}\vec{x}$$

which is equivalent to changes in kinetic energy (yielding the work energy theorem).

Power P is the rate of change $\mathrm{d}W/\mathrm{d}t$ of the work W, as the trajectory is extended by a position change $d\vec{x}$ in a time interval $\mathrm{d}t$:

$$\mathrm{d}W = \frac{\mathrm{d}W}{\mathrm{d}\vec{x}}\cdot\mathrm{d}\vec{x} = \vec{F}\cdot\mathrm{d}\vec{x}, \qquad \text{so} \qquad P = \frac{\mathrm{d}W}{\mathrm{d}t} = \frac{\mathrm{d}W}{\mathrm{d}\vec{x}}\cdot\frac{\mathrm{d}\vec{x}}{\mathrm{d}t} = \vec{F}\cdot\vec{v},$$

with $\vec{v} = \mathrm{d}\vec{x}/\mathrm{d}t$ the velocity.

Potential Energy

Instead of a force, often the mathematically related concept of a potential energy field can be used for convenience. For instance, the gravitational force acting upon an object can be seen as the action of the gravitational field that is present at the object's location. Restating mathematically the definition of energy (via the definition of work), a potential scalar field $U(\vec{r})$ is defined as that field whose gradient is equal and opposite to the force produced at every point:

$$\vec{F} = -\vec{\nabla}U.$$

Forces can be classified as conservative or nonconservative. Conservative forces are equivalent to the gradient of a potential while nonconservative forces are not.

Conservative Forces

A conservative force that acts on a closed system has an associated mechanical work that allows energy to convert only between kinetic or potential forms. This means that for a closed system, the net mechanical energy is conserved whenever a conservative force acts on the system. The force, therefore, is related directly to the difference in potential energy between two different locations in space, and can be considered to be an artifact of the potential field in the same way that the direction and amount of a flow of water can be considered to be an artifact of the contour map of the elevation of an area.

Conservative forces include gravity, the electromagnetic force, and the spring force. Each of these forces has models which are dependent on a position often given as a radial vector $\vec{r}$ emanating from spherically symmetric potentials. Examples of this follow:

For gravity:

$$\vec{F} = -\frac{Gm_1m_2\vec{r}}{r^3}$$

where G is the gravitational constant, and m_n is the mass of object n.

For electrostatic forces:

$$\vec{F} = \frac{q_1q_2\vec{r}}{4\pi\epsilon_0r^3}$$

where ϵ_0 is electric permittivity of free space, and q_n is the electric charge of object n.

For spring forces:

$$\vec{F} = -k\vec{r}$$

where k is the spring constant.

Nonconservative Forces

For certain physical scenarios, it is impossible to model forces as being due to gradient of potentials. This is often due to macrophysical considerations which yield forces as arising from a macroscopic statistical average of microstates. For example, friction is caused by the gradients of numerous electrostatic potentials between the atoms, but manifests as a force model which is independent of any macroscale position vector. Nonconservative forces other than friction include other contact forces, tension, compression, and drag. However, for any sufficiently detailed description, all these forces are the results of conservative ones since each of these macroscopic forces are the net results of the gradients of microscopic potentials.

The connection between macroscopic nonconservative forces and microscopic conservative forces is described by detailed treatment with statistical mechanics. In macroscopic closed systems, nonconservative forces act to change the internal energies of the system, and are often associated with the transfer of heat. According to the Second Law of Thermodynamics, nonconservative forces necessarily result in energy transformations within closed systems from ordered to more random conditions as entropy increases.

Types of Motion

Linear Motion

Linear motion (also called rectilinear motion) is motion along a straight line, and can therefore be described mathematically using only one spatial dimension. The linear motion can be of two types: uniform linear motion with constant velocity or zero acceleration; non uniform linear motion with variable velocity or non-zero acceleration. The motion of a particle (a point-like object) along a line can be described by its position x, which varies with t(time). An example of linear motion is an athlete running 100m along a straight track.

Linear motion is the most basic of all motion. According to Newton's first law of motion, objects that do not experience any net force will continue to move in a straight line with a constant velocity until they are subjected to a net force. Under everyday circumstances, external forces such as gravity and friction can cause an object to change the direction of its motion, so that its motion cannot be described as linear. One may compare linear motion to general motion. In general motion, a particle's position and velocity are described by vectors, which have a magnitude and direction. In linear motion, the directions of all the vectors describing the system are equal and constant which means the objects move along the same axis and do not change direction. The analysis of such systems may therefore be simplified by neglecting the direction components of the vectors involved and dealing only with the magnitude.

Neglecting the rotation and other motions of the Earth, an example of linear motion is the ball thrown straight up and falling back straight down.

Displacement

The motion in which all the particles of a body move through the same distance in the same time is called translatory motion. There are two types of translatory motions: rectilinear motion; curvilinear motion. Since linear motion is a motion in a single dimension, the distance travelled by an object in particular direction is the same as displacement. The SI unit of displacement is the metre. If x_1 is the initial position of an object and x_2 is the final position, then mathematically the displacement is given by:

$$\Delta x = x_2 - x_1$$

The equivalent of displacement in rotational motion is the angular displacement θ measured in radian. The displacement of an object cannot be greater than the distance. Consider a person travelling to work daily. Overall displacement when he returns home is zero, since the person ends up back where he started, but the distance travelled is clearly non zero.

Velocity

Velocity is defined as the rate of change of displacement with respect to time. The SI unit of velocity is ms^{-1} or metre per second.

Average Velocity

The average velocity is the ratio of total displacement Δx taken over time interval Δt. Mathematically, it is given by:

$$\mathrm{v}_{\mathrm{av}} = \frac{\Delta x}{\Delta t} = \frac{x_2 - x_1}{t_2 - t_1}$$

where:

t_1 is the time at which the object was at position x_1

t_2 is the time at which the object was at position x_2

Instantaneous Velocity

The instantaneous velocity can be found by differentiating the displacement with respect to time.

$$\mathrm{v} = \lim_{\Delta t \to 0} \frac{\Delta x}{\Delta t} = \frac{dx}{dt}$$

Speed

Speed is the absolute value of velocity i.e. speed is always positive. The unit of speed is metre per second. If $v$is the speed then,

$$v = |\mathrm{v}| = \left|\frac{dx}{dt}\right|$$

The magnitude of the instantaneous velocity is the instantaneous speed.

Acceleration

Acceleration is defined as the rate of change of velocity with respect to time. Acceleration is the second derivative of displacement i.e. acceleration can be found by differentiating position with respect to time twice or differentiating velocity with respect to time once. The SI unit of acceleration is ms^{-2} or metre per second squared.

If a_{av} is the average acceleration and $\Delta \mathrm{v} = \mathrm{v}_2 - \mathrm{v}_1$ is the average velocity over the time interval Δt then mathematically,

$$\mathrm{a}_{\mathrm{av}} = \frac{\Delta \mathrm{v}}{\Delta t} = \frac{\mathrm{v}_2 - \mathrm{v}_1}{t_2 - t_1}$$

The instantaneous acceleration is the limit of the ratio $\Delta \mathrm{v}$ and Δt as Δt approaches zero i.e.,

$$\mathrm{a} = \lim_{\Delta t \to 0} \frac{\Delta \mathrm{v}}{\Delta t} = \frac{d\mathrm{v}}{dt} = \frac{d^2 x}{dt^2}$$

Jerk

The rate of change of acceleration, the third derivative of displacement is known as jerk. The SI unit of jerk is ms^{-3}. In the UK jerk is also called as jolt.

Jounce

The rate of change of jerk, the fourth derivative of displacement is known as jounce. The SI unit of jounce is ms^{-4} which can be pronounced as *metres per quartic second.*

Equations of Kinematics

In case of constant acceleration, the four physical quantities acceleration, velocity, time and displacement can be related by using the Equations of motion

$$v = u + at$$

$$s = ut + \frac{1}{2}at^2$$

$$v^2 = u^2 + 2as$$

$$s = \tfrac{1}{2}(v + u)t$$

here,

u is the initial velocity

v is the final velocity

a is the acceleration

s is the displacement

t is the time

These relationships can be demonstrated graphically. The gradient of a line on a displacement time graph represents the velocity. The gradient of the velocity time graph gives the acceleration while the area under the velocity time graph gives the displacement. The area under an acceleration time graph gives the change in velocity.

Analogy between Linear and Rotational Motion

Table: *Analogy between Linear Motion and Rotational motion*

Linear motion	*Rotational motion*
Displacement = s	Angular displacement = θ
Velocity = v	Angular velocity = ω
Acceleration = a	Angular acceleration = α
Mass = m	Moment of Inertia = I
Force = $F = ma$	Torque = $T = I\alpha$

Rotation around a Fixed Axis

Rotation around a fixed axis is a special case of rotational motion. The fixed axis hypothesis excludes the possibility of a moving axis, and cannot describe such phenomena as wobbling or precession. According to Euler's rotation theorem, simultaneous rotation around more than one axis at the same time is impossible. If two rotations are forced at the same time, a new axis of rotation will appear. This article assumes that the rotation is also stable, such that no torque is required to keep it going. The kinematics and dynamics of rotation around a fixed axis of a rigid body are mathematically much simpler than those for free rotation of a rigid body; they are entirely analogous to those of linear motion along a single fixed direction, which is not true for *free rotation of a rigid body*. The expressions for the kinetic energy of the object, and for the forces on the parts of the object, are also simpler for rotation around a fixed axis, than for general rotational motion. For these reasons, rotation around a fixed axis is typically taught in introductory physics courses after students have mastered linear motion; the full generality of rotational motion is not usually taught in introductory physics classes.

Translation and Rotation

A *rigid body* is an object of finite extent in which all the distances between the component particles are constant. No truly rigid body exists; external forces can deform any solid. For our purposes, then, a rigid body is a solid which requires large forces to deform it appreciably.

A change in the position of a particle in three-dimensional space can be completely specified by three coordinates. A change in the position of a rigid body is more complicated to describe. It can be regarded as a combination of two distinct types of motion: translational motion and rotational motion.

Purely *translational motion* occurs when every particle of the body has the same instantaneous velocity as every other particle; then the path traced out by any particle is exactly parallel to the path traced out by every other particle in the body. Under translational motion, the change in the position of a rigid body is specified completely by three coordinates such as x, y, and z giving the displacement of any point, such as the centre of mass, fixed to the rigid body.

Purely *rotational motion* occurs if every particle in the body moves in a circle about a single line. This line is called the axis of rotation. Then the radius vectors from the axis to all particles undergo the same angular displacement in the same time. The axis of rotation need not go through the body. In general, any rotation can be specified completely by the three angular displacements with respect to the rectangular-coordinate axes x, y, and z. Any change in the position of the rigid body is thus completely described by three translational and three rotational coordinates. Any displacement of a rigid body may be arrived at by first subjecting the body to a displacement followed by a rotation, or conversely, to a rotation followed by a displacement. We already know that for any collection of particles—whether at rest with respect to one another, as in a rigid body, or in relative motion, like the exploding fragments of a shell, the acceleration of the centre of mass is given by

$$F_{net} = Ma_{cm}$$

where M is the total mass of the system and a_{cm} is the acceleration of the centre of mass. There remains the matter of describing the rotation of the body about the centre of mass and relating it to the external forces acting on the body. The kinematics and dynamics of *rotational motion around a single axis* resemble the kinematics and dynamics of translational motion; rotational motion around a single axis even has a work-energy theorem analogous to that of particle dynamics.

Kinematics

Angular Displacement: A particle moves in a circle of radius r. Having moved an arc length s, its angular position is θ relative to its original position, where $\theta = \frac{s}{r}$.

In mathematics and physics it is usual to use the natural unit radians rather than degrees or revolutions. Units are converted as follows:

$$1 \text{ rev} = 360^\circ = 2\pi \text{ rad, and}$$

$$1 \text{ rad} = 180^\circ / \pi \approx 57.3^\circ.$$

An angular displacement is a change in angular position:

$$\Delta\theta = \theta_2 - \theta_1,$$

where $\Delta\theta$ is the angular displacement, θ_1 is the initial angular position and θ_2 is the final angular position.

Angular Speed and Angular Velocity

Angular velocity is the change in angular displacement per unit time. The symbol for angular velocity is ω and the units are typically rad s^{-1}. Angular speed is the magnitude of angular velocity.

$$\bar{\omega} = \frac{\Delta\theta}{\Delta t} = \frac{\theta_2 - \theta_1}{t_2 - t_1}.$$

The instantaneous angular velocity is given by

$$\omega(t) = \frac{d\theta}{dt}$$

Using the formula for angular position and letting $v = \frac{ds}{dt}$, we have also

$$\omega = \frac{d\theta}{dt} = \frac{v}{r},$$

where v is the translational speed of the particle.

Angular velocity and frequency are related by

$$\omega = 2\pi f \ .$$

Angular Acceleration

A changing angular velocity indicates the presence of an angular acceleration in rigid body, typically measured in rad s^{-2}. The average angular acceleration $\overline{\alpha}$ over a time interval Δt is given by

$$\overline{\alpha} = \frac{\Delta\omega}{\Delta t} = \frac{\omega_2 - \omega_1}{t_2 - t_1}.$$

The instantaneous acceleration $\alpha(t)$ is given by

$$\alpha(t) = \frac{d\omega}{dt} = \frac{d^2\theta}{dt^2}.$$

Thus, the angular acceleration is the rate of change of the angular velocity, just as acceleration is the rate of change of velocity. The translational acceleration of a point on the object rotating is given by

$$a = r\alpha,$$

where r is the radius or distance from the axis of rotation. This is also the tangential component of acceleration: it is tangential to the direction of motion of the point. If this component is 0, the motion is uniform circular motion, and the velocity changes in direction only. The radial acceleration (perpendicular to direction of motion) is given by

$$a_R = \frac{v^2}{r} = \omega^2 r \ .$$

It is directed towards the centre of the rotational motion, and is often called the *centripetal acceleration.* The angular acceleration is caused by the torque, which can have a positive or negative value in accordance with the convention of positive and negative angular frequency. The ratio of torque and angular acceleration (how difficult it is to start, stop, or otherwise change rotation) is given by the moment of inertia: $T = Ia$.

Equations of Kinematics

When the angular acceleration is constant, the five quantities angular displacement θ, initial angular velocity ω_i, final angular velocity

ω_f, angular acceleration α, and time t can be related by four equations of kinematics:

$$\omega_f = \omega_i + \alpha t$$

$$\theta = \omega_i t + \frac{1}{2}\alpha t^2$$

$$\omega_f^2 = \omega_i^2 + 2\alpha\theta$$

$$\theta = \tfrac{1}{2}\left(\omega_f + \omega_i\right)t$$

Dynamics

Moment of Inertia: The moment of inertia of an object, symbolized by I, is a measure of the object's resistance to changes to its rotation. The moment of inertia is measured in kilogram metre² (kg m²). It depends on the object's mass: increasing the mass of an object increases the moment of inertia. It also depends on the distribution of the mass: distributing the mass further from the centre of rotation increases the moment of inertia by a greater degree. For a single particle of mass m a distance r from the axis of rotation, the moment of inertia is given by

$$I = mr^2.$$

Torque

Torque τ is the twisting effect of a force F applied to a rotating object which is at position r from its axis of rotation. Mathematically,

$$\tau = \mathrm{r} \times \mathrm{F},$$

where × denotes the cross product. A net torque acting upon an object will produce an angular acceleration of the object according to

$$\tau = I\alpha,$$

just as F = ma in linear dynamics. The work done by a torque acting on an object equals the magnitude of the torque times the angle through which the torque is applied:

$$W = T\theta.$$

The power of a torque is equal to the work done by the torque per unit time, hence:

$$P = T\omega.$$

Angular Momentum

The angular momentum L is a measure of the difficulty of bringing a rotating object to rest. It is given by

$$L = r \times p.$$

Angular momentum is related to angular velocity by

$$L = I\omega,$$

just as $p = mv$ in linear dynamics.

Torque and angular momentum are related according to

$$\tau = \frac{dL}{dt},$$

just as $F = dp/dt$ in linear dynamics. In the absence of an external torque, the angular momentum of a body remains constant. The conservation of angular momentum is notably demonstrated in figure skating: when pulling the arms closer to the body during a spin, the moment of inertia is decreased, and so the angular velocity is increased.

Kinetic energy

The kinetic energy K_{rot} due to the rotation of the body is given by

$$K_{rot} = \tfrac{1}{2} I\omega^2,$$

just as $K_{trans} = {}^1/_2 mv^2$ in linear dynamics.

Vector Expression

The above development is a special case of general rotational motion. In the general case, angular displacement, angular velocity, angular acceleration and torque are considered to be vectors. An angular displacement is considered to be a vector, pointing along the axis, of magnitude equal to that of $\Delta\theta$. A right-hand rule is used to find which way it points along the axis; if the fingers of the right hand are curled to point in the way that the object has rotated, then the thumb of the right hand points in the direction of the vector.

The angular velocity vector also points along the axis of rotation in the same way as the angular displacements it causes. If a disk spins counterclockwise as seen from above, its angular velocity vector points upwards. Similarly, the angular acceleration vector points along the axis of rotation in the same direction that the angular velocity would point if the angular acceleration were maintained for a long time.

The torque vector points along the axis around which the torque tends to cause rotation. To maintain rotation around a fixed axis, the total torque vector has to be along the axis, so that it only changes the magnitude and not the direction of the angular velocity vector. In the case of a hinge, only the component of the torque vector along the axis has effect on the rotation, other forces and torques are compensated by the structure.

Examples and Applications

Constant Angular Speed: The simplest case of rotation around a fixed axis is that of constant angular speed. Then the total torque is zero. For the example of the Earth rotating around its axis, there is very little friction. For a fan, the motor applies a torque to compensate for friction. The angle of rotation is a linear function of time, which modulo 360° is a periodic function.

An example of this is the two-body problem with circular orbits.

Centripetal Force

Internal tensile stress provides the centripetal force that keeps a spinning object together. A rigid body model neglects the accompanying strain. If the body is not rigid this strain will cause it to change shape. This is expressed as the object changing shape due to the "centrifugal force".

Celestial bodies rotating about each other often have elliptic orbits. The special case of circular orbits is an example of a rotation around a fixed axis: this axis is the line through the centre of mass perpendicular to the plane of motion. The centripetal force is provided by gravity. This usually also applies for a spinning celestial body, so it need not be solid to keep together, unless the angular speed is too high in relation to its density. (It will, however, tend to become oblate.) For example, a spinning celestial body of water must take at least 3 hours and 18 minutes to rotate, regardless of size, or the water will separate. If the density of the fluid is higher the time can be less.

Circular Motion

In physics, circular motion is a movement of an object along the circumference of a circle or rotation along a circular path. It can be uniform, with constant angular rate of rotation (and constant speed), or non-uniform with a changing rate of rotation. The rotation around a fixed axis of a three-dimensional body involves circular motion of its parts. The equations of motion describe the movement of the centre of mass of a body.

Examples of circular motion include: an artificial satellite orbiting the Earth at constant height, a stone which is tied to a rope and is being swung in circles, a car turning through a curve in a race track, an electron moving perpendicular to a uniform magnetic field, and a gear turning inside a mechanism. Since the object's velocity vector is constantly changing direction, the moving object is undergoing acceleration by a centripetal force in the direction of the centre of rotation. Without this acceleration, the object would move in a straight line, according to Newton's laws of motion.

Uniform Circular Motion

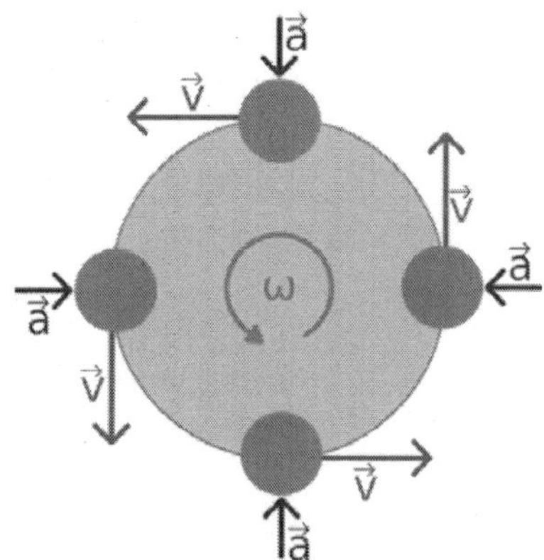

Figure: *Velocity v and acceleration a in uniform circular motion at angular rate ω; the speed is constant, but the velocity is always tangent to the orbit; the acceleration has constant magnitude, but always points toward the centre of rotation*

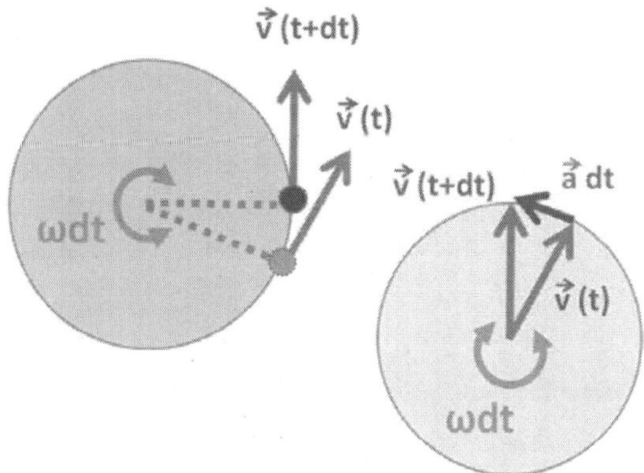

Figure: *The velocity vectors at time* t *and time* t + dt *are moved from the orbit on the left to new positions where their tails coincide, on the right. Because the velocity is fixed in magnitude at* v = r ω, *the velocity vectors also sweep out a circular path at angular rate* ω. *As* dt → *0, the acceleration vector a becomes perpendicular to v, which means it points toward the centre of the orbit in the circle on the left. Angle* ω dt *is the very small angle between the two velocities and tends to zero as* dt→ *0*

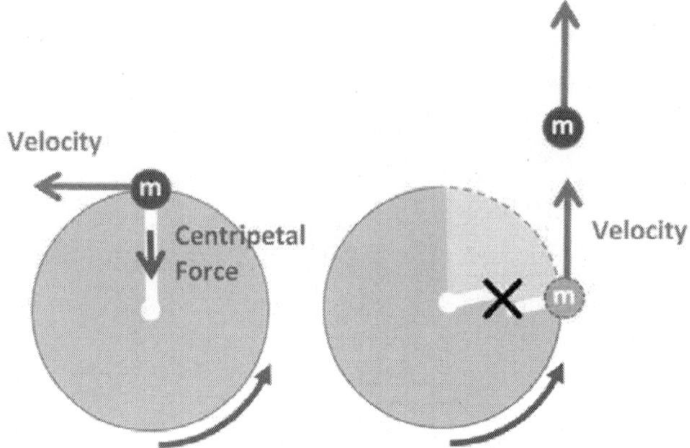

Figure: *(Left) Ball in circular motion – rope provides centripetal force to keep ball in circle (Right) Rope is cut and ball continues in straight line with velocity at the time of cutting the rope, in accord with Newton's law of inertia, because centripetal force is no longer there*

In physics, uniform circular motion describes the motion of a body traversing a circular path at constant speed. The distance of the body from the axis of rotation remains constant at all times. Though the body's speed is constant, its velocity is not constant: velocity, a vector quantity, depends on both the body's speed and its direction of travel. This changing velocity indicates the presence of an acceleration; this centripetal acceleration is of constant magnitude and directed at all times towards the axis of rotation. This acceleration is, in turn, produced by a centripetal force which is also constant in magnitude and directed towards the axis of rotation.

In the case of rotation around a fixed axis of a rigid body that is not negligibly small compared to the radius of the path, each particle of the body describes a uniform circular motion with the same angular velocity, but with velocity and acceleration varying with the position with respect to the axis.

Formulas for Uniform Circular Motion

For motion in a circle of radius r, the circumference of the circle is $C = 2\pi\ r$. If the period for one rotation is T, the angular rate of rotation, also known as angular velocity, ω is:

$$\omega = \frac{2\pi}{T} \text{ and the units are radians/sec}$$

The speed of the object travelling the circle is:

$$v = \frac{2\pi r}{T} - \omega r$$

The angle θ swept out in a time *t* is:

$$\theta = 2\pi \frac{t}{T} = \omega t$$

The acceleration due to change in the direction is:

$$a = \frac{v^2}{r} = \omega^2 r$$

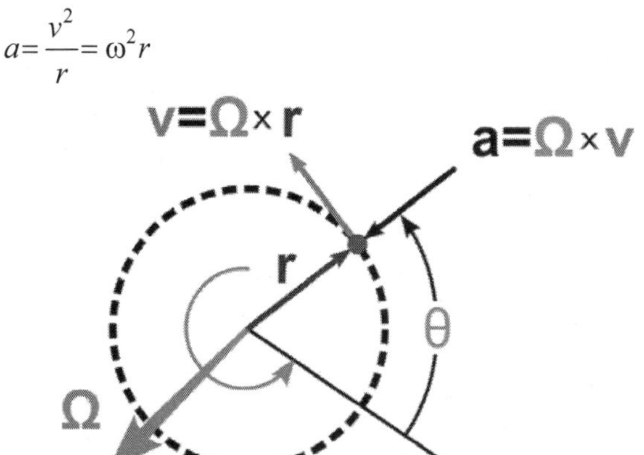

Figure: *Vector relationships for uniform circular motion; vector Ω representing the rotation is normal to the plane of the orbit.*

The vector relationships are shown in Figure. The axis of rotation is shown as a vector Ω perpendicular to the plane of the orbit and with a magnitude $\omega = d\theta / dt$. The direction of Ω is chosen using the right-hand rule. With this convention for depicting rotation, the velocity is given by a vector cross product as

$$\mathrm{v} = \Omega \times \mathrm{r},$$

which is a vector perpendicular to both Ω and r (*t*), tangential to the orbit, and of magnitude ω *r*. Likewise, the acceleration is given by

$$\mathrm{a} = \Omega \times \mathrm{v} = \Omega \times (\Omega \times \mathrm{r}),$$

which is a vector perpendicular to both Ω and v (*t*) of magnitude $\omega\ |\mathrm{v}| = \omega^2 r$ and directed exactly opposite to r (*t*).

In the simplest case the speed, mass and radius are constant.

Consider a body of one kilogram, moving in a circle of radius one metre, with an angular velocity of one radian per second.

- The speed is one metre per second.
- The inward acceleration is one metre per square second[v^2/r]
- It is subject to a centripetal force of one kilogram metre per square second, which is one newton.
- The momentum of the body is one $\mathrm{kg \cdot m \cdot s^{-1}}$.

- The moment of inertia is one kg·m².
- The angular momentum is one kg·m²·s⁻¹.
- The kinetic energy is 1/2 joule.
- The circumference of the orbit is 2π (~ 6.283) metres.
- The period of the motion is 2π seconds per turn.
- The frequency is $(2\pi)^{-1}$hertz.

In Polar Coordinates

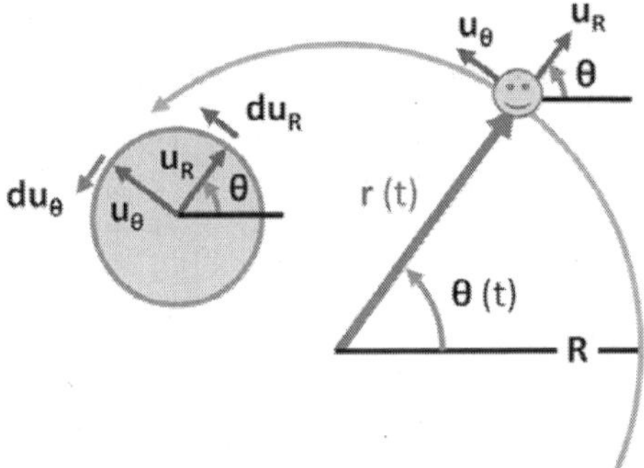

Figure: *Polar coordinates for circular trajectory. On the left is a unit circle showing the changes* $d\hat{u}_R$ *and* $d\hat{u}_\theta$ *in the unit vectors* $\hat{u}_R$ *and* $\hat{u}_\theta$ *for a small increment* $d\theta$ *in angle* θ.

During circular motion the body moves on a curve that can be described in polar coordinate system as a fixed distance R from the centre of the orbit taken as origin, oriented at an angle $\theta\,(t)$ from some reference direction. The displacement *vector* $\vec{r}$ is the radial vector from the origin to the particle location:

$$\vec{r} = R\hat{u}_R(t),$$

where $\hat{u}_R(t)$ is the unit vector parallel to the radius vector at time t and pointing away from the origin. It is convenient to introduce the unit vector orthogonal to $\hat{u}_R$ as well, namely $\hat{u}_\theta$. It is customary to orient $\hat{u}_\theta$ to point in the direction of travel along the orbit.

The velocity is the time derivative of the displacement:

$$\vec{v} = \frac{d}{dt}\vec{r}(t) = \frac{dR}{dt}\hat{u}_R + R\frac{d\hat{u}_R}{dt}.$$

Because the radius of the circle is constant, the radial component of the velocity is zero. The unit vector $\hat{u}_R$ has a time-invariant magnitude of unity, so as time varies its tip always lies on a circle of unit radius, with an angle 0 the same as the angle of $\vec{r}(t)$. If the particle displacement rotates through an angle $d\theta$ in time dt, so does $\hat{u}_R$, describing an arc on the unit circle of magnitude $d\theta$. Hence:

$$\frac{d\hat{u}_R}{dt} = \frac{d\theta}{dt}\hat{u}_\theta ,$$

where the direction of the change must be perpendicular to $\hat{u}_R$ (or, in other words, along $\hat{u}_\theta$) because any change $d\ \hat{u}_R$ in the direction of $\hat{u}_R$ would change the size of $\hat{u}_R$. The sign is positive, because an increase in $d\theta$ implies the object and have moved in the direction of. Hence the velocity becomes:

$$\vec{v} = \frac{d}{dt}\vec{r}(t) = R\frac{d\hat{u}_R}{dt} = R\frac{d\theta}{dt}\hat{u}_\theta = R\omega\hat{u}_\theta .$$

The acceleration of the body can also be broken into radial and tangential components. The acceleration is the time derivative of the velocity:

$$\vec{a} = \frac{d}{dt}\vec{v} = \frac{d}{dt}\left(R\,\omega\,\hat{u}_\theta\right).$$

$$= R\left(\frac{d\omega}{dt}\hat{u}_\theta + \omega\frac{d\hat{u}_\theta}{dt}\right).$$

The time derivative of $\hat{u}_\theta$ is found the same way as for $\hat{u}_R$. Again, $\hat{u}_\theta$ is a unit vector and its tip traces a unit circle with an angle that is $\pi/2 + \theta$. Hence, an increase in angle $d\theta$ by $\vec{r}(t)$ implies $\hat{u}_\theta$ traces an arc of magnitude $d\theta$, and as $\hat{u}_\theta$ is orthogonal to $\hat{u}_R$, we have:

$$\frac{d\hat{u}_\theta}{dt} = -\frac{d\theta}{dt}\hat{u}_R = -\omega\hat{u}_R ,$$

where a negative sign is necessary to keep $\hat{u}_\theta$ orthogonal to $\hat{u}_R$. (Otherwise, the angle between $\hat{u}_\theta$ and $\hat{u}_R$ would *decrease* with increase in $d\theta$.) Consequently the acceleration is:

$$\vec{a} = R\left(\frac{d\omega}{dt}\hat{u}_\theta + \omega\frac{d\hat{u}_\theta}{dt}\right)$$

$$= R\frac{d\omega}{dt}\hat{u}_\theta - \omega^2 R\,\hat{u}_R .$$

The centripetal acceleration is the radial component, which is directed radially inward:

$$\vec{a}_R = -\omega^2 R\hat{u}_R ,$$

while the tangential component changes the magnitude of the velocity:

$$\vec{a}_\theta = R\frac{d\omega}{dt}\hat{u}_\theta = \frac{dR\omega}{dt}\hat{u}_\theta = \frac{d|\vec{v}|}{dt}\hat{u}_\theta .$$

Using Complex Numbers

Circular motion can be described using complex numbers. Let the x axis be the real axis and the y axis be the imaginary axis. The position of the body can then be given as z, a complex "vector":

$$z = x + iy = R(\cos\theta + i\sin\theta) = Re^{i\theta} ,$$

where i is the imaginary unit, and

$$\theta = \theta(t) ,$$

is the angle of the complex vector with the real axis and is a function of time t. Since the radius is constant:

$$\dot{R} = \ddot{R} = 0 ,$$

where a *dot* indicates time differentiation. With this notation the velocity becomes:

$$v = \dot{z} = \frac{d(Re^{i\theta})}{dt} = R\frac{d\theta}{dt}\frac{d(e^{i\theta})}{d\theta} = iR\dot{\theta}e^{i\theta} = i\omega \cdot Re^{i\theta} = i\omega z$$

and the acceleration becomes:

$$a = \dot{v} = i\dot{\omega}z + i\omega\dot{z} = (i\dot{\omega} - \omega^2)z$$

$$= \left(i\dot{\omega} - \omega^2\right)Re^{i\theta}$$

$$= -\omega^2 Re^{i\theta} + \dot{\omega}e^{i\frac{\pi}{2}}Re^{i\theta} .$$

The first term is opposite in direction to the displacement vector and the second is perpendicular to it, just like the earlier results shown before.

Velocity

Figure illustrates velocity and acceleration vectors for uniform motion at four different points in the orbit. Because the velocity v is tangent to the circular path, no two velocities point in the same direction. Although the object has a constant *speed*, its *direction* is always changing. This change in velocity is caused by an acceleration a, whose magnitude is (like that of the velocity) held constant, but whose direction also is always changing. The acceleration points radially inwards (centripetally) and is perpendicular to the velocity. This acceleration is known as centripetal acceleration.

For a path of radius r, when an angle θ is swept out, the distance travelled on the periphery of the orbit is $s = r\theta$. Therefore, the speed of travel around the orbit is

$$v = r\frac{d\theta}{dt} = r\omega,$$

where the angular rate of rotation is ω. (By rearrangement, $\omega = v/r$.) Thus, v is a constant, and the velocity vector v also rotates with constant magnitude v, at the same angular rate ω.

Non-uniform

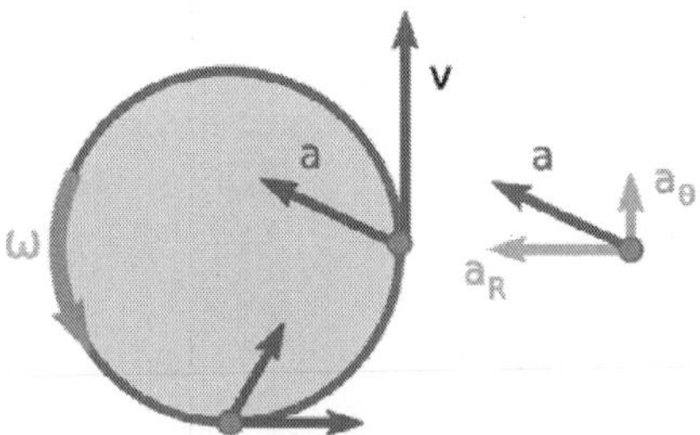

Non-uniform circular motion is any case in which an object moving in a circular path has a varying speed. The tangential acceleration is non-zero; the speed is changing.

Since there is a non-zero tangential acceleration, there are forces that act on an object in addition to its centripetal force (composed of the mass and radial acceleration). These forces include weight, normal force, and friction.

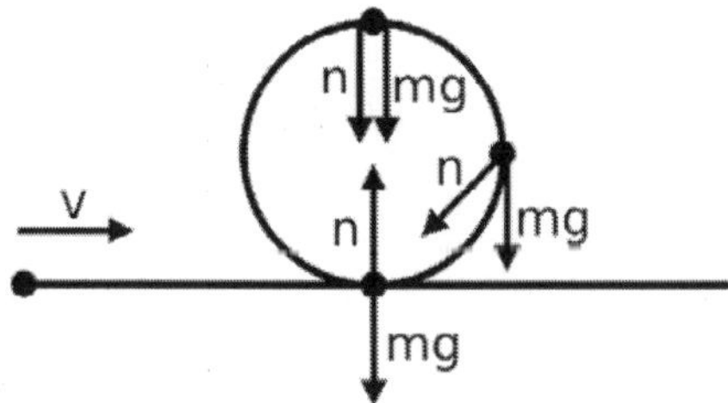

In non-uniform circular motion, normal force does not always point in the opposite direction of weight. Here is an example with an object travelling in a straight path then loops a loop back into a straight path again.

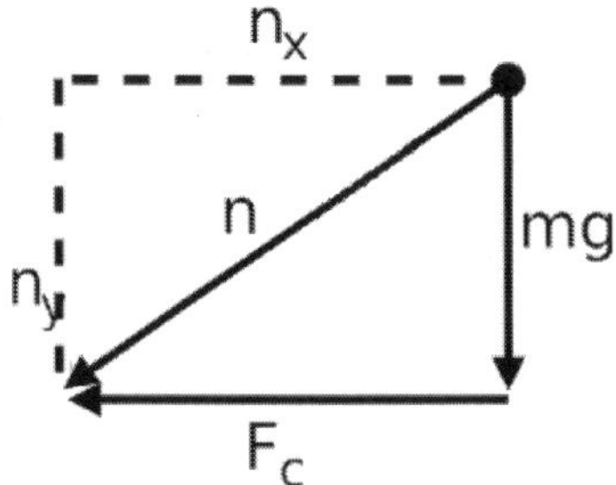

This diagram shows the normal force pointing in other directions rather than opposite to the weight force. The normal force is actually the sum of the radial and tangential forces that help to counteract the weight force and contribute to the centripetal force. The horizontal component of normal force is what contributes to the centripetal force. The vertical component of the normal force is what counteracts the weight of the object.

In non-uniform circular motion, normal force and weight may point in the same direction. Both forces can point down, yet the object will remain in a circular path without falling straight down. First let's see why normal force can point down in the first place. In the first diagram, let's say the object is a person sitting inside a plane, the two forces point down only when it reaches the top of the circle. The reason for this is that the normal force is the sum of the weight and centripetal force. Since both weight and centripetal force points down at the top of the circle, normal force will point down as well. From a logical standpoint, a person who is travelling in the plane will be upside down at the top of the circle. At that moment, the person's seat is actually pushing down on the person, which is the normal force.

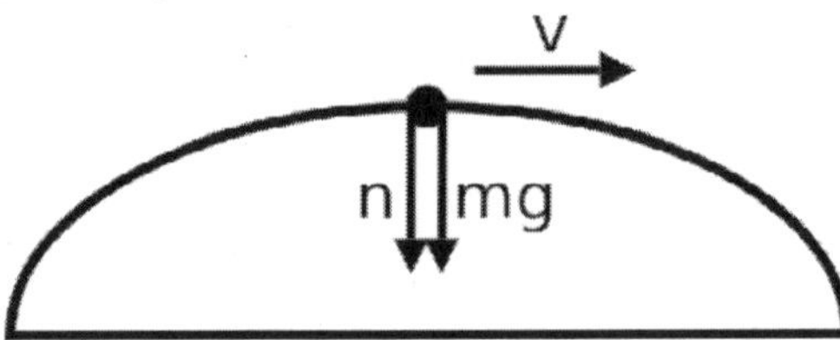

The reason why the object does not fall down when subjected to only downward forces is a simple one. Think about what keeps an object up after it is thrown. Once an object is thrown into the air, there is only the downward force of earth's gravity that acts on the object. That does not mean that once an object is thrown in the air, it will fall instantly. What keeps that object up in the air is its velocity. The first of Newton's laws of motion states that an object's inertia keeps it in motion, and since the object in the air has a velocity, it will tend to keep moving in that direction.

Applications

Solving applications dealing with non-uniform circular motion involves force analysis. With uniform circular motion, the only force acting upon an object travelling in a circle is the centripetal force. In non-uniform circular motion, there are additional forces acting on the object due to a non-zero tangential acceleration. Although there are additional forces acting upon the object, the sum of all the forces acting on the object will have to equal to the centripetal force.

$$F_{net} = ma$$

$$F_{net} = ma_r$$

$$F_{net} = mv^2 / r$$

$$F_{net} = F_c$$

Radial acceleration is used when calculating the total force. Tangential acceleration is not used in calculating total force because it is not responsible for keeping the object in a circular path. The only acceleration responsible for keeping an object moving in a circle is the radial acceleration. Since the sum of all forces is the centripetal force, drawing centripetal force into a free body diagram is not necessary and usually not recommended.

Using, $F_{net} = F_c$ we can draw free body diagrams to list all the forces acting on an object then set it equal to F_c . Afterwards, we can solve for what ever is unknown (this can be mass, velocity, radius of curvature, coefficient of friction, normal force, etc.). For example, the visual above showing an object at the top of a semicircle would be expressed as $F_c = (n + mg)$.

In uniform circular motion, total acceleration of an object in a circular path is equal to the radial acceleration. Due to the presence of tangential acceleration in non uniform circular motion, that does not hold true any more. To find the total acceleration of an object in non uniform circular, find the vector sum of the tangential acceleration and the radial acceleration.

$$\sqrt{a_r^2 + a_t^2} = a$$

Radial acceleration is still equal to v^2 / r . Tangential acceleration is simply the derivative of the velocity at any given point: $a_t = dv / dt$. This root sum of squares of separate radial and tangential accelerations is only correct for circular motion; for general motion within a plane with polar coordinates (r, θ) , the Coriolis term $a_c = 2(dr / dt)(d\theta / dt)$ should be added to a_t , whereas radial acceleration then becomes $a_r = -v^2 / r + d^2r / dt^2$.

Perpetual Motion

Perpetual motion describes "motion that continues indefinitely without any external source of energy; impossible in practice because of friction." It can also be described as "the motion of a hypothetical machine which, once activated, would run forever unless subject to an external force or to wear". There is a scientific consensus that perpetual motion is impossible, as it would violate the first or second law of thermodynamics.

Cases of apparent perpetual motion can exist in nature, but either are not truly perpetual or cannot be used to do work without changing the nature of the motion. For example, the motion or rotation of celestial bodies such as planets may appear perpetual, but actually is slowing down due to energy loss from friction with intersteller gas, tides, and gravitational radiation and will eventually stop. The flow of electric current in a superconducting loop may be perpetual and could be used as an energy storage medium, but as with a battery the amount of energy stored is limited so it could not serve as a perpetual *source* of energy.

Machines which extract energy from seemingly perpetual sources—such as ocean currents—are capable of moving "perpetually" (for as long as that energy source itself endures), but they are not considered to be perpetual motion machines because they are consuming energy from an external source and are not isolated systems (in reality, no system can ever be a fully isolated system). Similarly, machines which comply with both laws of thermodynamics but access energy from obscure sources are sometimes referred to as perpetual motion machines, although they also do not meet the standard criteria for the name. Despite the fact that successful isolated system perpetual motion devices are physically impossible in terms of the current understanding of the laws of physics, the pursuit of perpetual motion remains popular.

Basic Principles

There is a scientific consensus that perpetual motion in an isolated system violates either the first law of thermodynamics, the second law of thermodynamics, or both. The first law of thermodynamics is essentially a statement of conservation of energy. The second law can be phrased in several different ways, the most intuitive of which is that heat flows spontaneously from hotter to colder places; the most well known statement is that entropy tends to increase, or at the least stay the same; another statement is that no heat engine (an engine which produces work while moving heat from a high temperature to a low temperature) can be more efficient than a Carnot heat engine.

In other words:

1. In any isolated system, one cannot create new energy (first law of thermodynamics)

2. The output power of heat engines is always smaller than the input heating power. The rest of the energy is removed as heat at ambient temperature. The efficiency (this is the produced power divided by the input heating power) has a maximum, given by the Carnot efficiency. It is always lower than one
3. The efficiency of real heat engines is even lower than the Carnot efficiency due to irreversible processes.

The statements 2 and 3 only apply to heat engines. Other types of engines, which convert e.g. mechanical into electromagnetic energy, can, in principle, operate with 100% efficiency.

Machines which comply with both laws of thermodynamics by accessing energy from unconventional sources are sometimes referred to as perpetual motion machines, although they do not meet the standard criteria for the name. By way of example, clocks and other low-power machines, such as Cox's timepiece, have been designed to run on the differences in barometric pressure or temperature between night and day. These machines have a source of energy, albeit one which is not readily apparent so that they only seem to violate the laws of thermodynamics.

Machines which extract energy from seemingly perpetual sources - such as ocean currents - are indeed capable of moving "perpetually" until that energy source runs down. They are not considered to be perpetual motion machines because they are consuming energy from an external source and are not isolated systems.

Classification

One classification of perpetual motion machines refers to the particular law of thermodynamics the machines purport to violate:

- A perpetual motion machine of the first kind produces work without the input of energy. It thus violates the first law of thermodynamics: the law of conservation of energy.
- A perpetual motion machine of the second kind is a machine which spontaneously converts thermal energy into mechanical work. When the thermal energy is equivalent to the work done, this does not violate the law of conservation of energy. However it does violate the more subtle second law of thermodynamics. The signature of a perpetual motion machine of the second kind is that there is only one heat reservoir involved, which is being spontaneously cooled without involving a transfer of heat to a cooler reservoir. This conversion of heat into useful work, without

any side effect, is impossible, according to the second law of thermodynamics.

- A more obscure category is a perpetual motion machine of the third kind, usually (but not always) defined as one that completely eliminates friction and other dissipative forces, to maintain motion forever (due to its mass inertia). *Third* in this case refers solely to the position in the above classification scheme, not the third law of thermodynamics. Although it is impossible to make such a machine, as dissipation can never be 100% eliminated in a mechanical system, it is nevertheless possible to get very close to this ideal. Such a machine would not serve as a source of energy but would have utility as a perpetual energy storage device.

Use of the Term "impossible" and Perpetual Motion

"Epistemic impossibility" describes things which absolutely cannot occur within our *current* formulation of the physical laws. This interpretation of the word "impossible" is what is intended in discussions of the impossibility of perpetual motion in a closed system.

The conservation laws are particularly robust from a mathematical perspective. Noether's theorem, which was proven mathematically in 1915, states that any conservation law can be derived from a corresponding continuous symmetry of the action of a physical system. For example, if the true laws of physics remain invariant over time then the conservation of energy follows. On the other hand, if the conservation laws are invalid, then the foundations of physics would need to change.

Scientific investigations as to whether the laws of physics are invariant over time use telescopes to examine the universe in the distant past to discover, to the limits of our measurements, whether ancient stars were identical to stars today. Combining different measurements such as spectroscopy, direct measurement of the speed of light in the past and similar measurements demonstrates that physics has remained substantially the same, if not identical, for all of observable history spanning billions of years.

The principles of thermodynamics are so well established, both theoretically and experimentally, that proposals for perpetual motion machines are universally met with disbelief on the part of physicists. Any proposed perpetual motion design offers a potentially instructive challenge to physicists: one is almost completely certain that it cannot work, so one must explain *how* it fails to work. The difficulty (and the value) of such an exercise depends on the subtlety of the proposal; the best ones tend to arise from physicists' own thought experiments and often shed light upon certain aspects of physics. So, for example, the thought experiment of a Brownian ratchet as a perpetual motion machine was first discussed by Gabriel Lippmann in 1900 but it was not until 1912 that Marian Smoluchowski gave an adequate explanation for why it cannot work. However, during that twelve-year period scientists did not believe that the machine was possible. They were merely unaware of the exact mechanism by which it would inevitably fail.

The law that entropy always increases, holds, I think, the supreme position among the laws of Nature. If someone points out to you that your pet theory of the universe is in disagreement with Maxwell's equations — then so much the worse for Maxwell's equations. If it is found to be contradicted by observation — well, these experimentalists do bungle things sometimes. But if your theory is found to be against the second law of thermodynamics I can give you no hope; there is nothing for it but to collapse in deepest humiliation.

—Sir Arthur Stanley Eddington, *The Nature of the Physical World* (1927)

In the mid 19th-century Henry Dircks investigated the history of perpetual motion experiments, writing a vitriolic attack on those who continued to attempt what he believed to be impossible:

> *"There is something lamentable, degrading, and almost insane in pursuing the visionary schemes of past ages with dogged determination, in paths of learning which have been investigated by superior minds, and with which such adventurous persons are totally unacquainted. The history of Perpetual Motion is a history of the fool-hardiness of either half-learned, or totally ignorant persons." —Henry Dircks, Perpetuum Mobile: Or, A History of the Search for Self-motive (1861)*

Techniques

Some common ideas reoccur repeatedly in perpetual motion machine designs. Many ideas that continue to appear today were stated as early as 1670 by John Wilkins, Bishop of Chester and an official of the Royal Society. He outlined three potential sources of power for a perpetual motion machine, "Chymical Extractions", "Magnetical Virtues" and "the Natural Affection of Gravity".

The seemingly mysterious ability of magnets to influence motion at a distance without any apparent energy source has long appealed to inventors. One of the earliest examples of a magnetic motor was proposed by Wilkins and has been widely copied since: it consists of a ramp with a magnet at the top, which pulled a metal ball up the ramp. Near the magnet was a small hole that was supposed to allow the ball to drop under the ramp and return to the bottom, where a flap allowed it to return to the top again. The device simply could not work: any magnet strong enough to pull the ball up the ramp would necessarily be too powerful to allow it to drop through the hole. Faced with this problem, more modern versions typically use a series of ramps and magnets, positioned so the ball is to be handed off from one magnet to another as it moves. The problem remains the same.

Gravity also acts at a distance, without an apparent energy source. But to get energy out of a gravitational field (for instance, by dropping a heavy object, producing kinetic energy as it falls) one has to put energy in (for instance, by lifting the object up), and some energy is always dissipated in the process. A typical application of gravity in a perpetual motion machine is Bhaskara's wheel in the 12th century, whose key idea is itself a recurring theme, often called the overbalanced wheel: Moving weights are attached to a wheel in such a way that they fall to a position further from the wheel's centre for one half of the wheel's rotation, and closer to the centre for the other half. Since weights further from the centre apply a greater torque, the result is (or would be, if such a device worked) that the wheel rotates forever. The moving weights may be hammers on pivoted arms, or rolling balls, or mercury in tubes; the principle is the same.

Yet another theoretical machine involves a frictionless environment for motion. This involves the use of diamagnetic or electromagnet levitation to float an object. This is done in a vacuum to eliminate air friction and friction from an axle. The levitated object is then free to rotate around its centre of gravity without interference. However, this machine has no practical purpose because the rotated object cannot do any work as work requires the levitated object to cause motion in other objects, bringing friction into the problem. Furthermore, a *perfect* vacuum is an unattainable goal since both the container and the object itself would slowly vaporize, thereby degrading the vacuum.

To extract work from heat, thus producing a perpetual motion machine of the second kind, the most common approach (dating back at least to Maxwell's demon) is *unidirectionality*. Only molecules moving fast enough and in the right direction are allowed through the demon's trap door. In a Brownian ratchet, forces tending to turn the ratchet one way are able to do so while forces in the other direction are not.

A diode in a heat bath allows through currents in one direction and not the other. These schemes typically fail in two ways: either maintaining the unidirectionality costs energy (It would require Maxwell's demon to perform more thermodynamic work to gauge the speed of the molecules than the amount of energy gained by the difference of temperature caused) or the unidirectionality is an illusion and occasional big violations make up for the frequent small non-violations (the Brownian ratchet will be subject to internal Brownian forces and therefore will sometimes turn the wrong way).

Buoyancy is another frequently-misunderstood phenomenon. Some proposed perpetual-motion machines miss the fact that to push a volume of air down in a fluid takes the same work as to raise a corresponding volume of fluid up against gravity. These types of machines may involve two chambers with pistons, and a mechanism to squeeze the air out of the top chamber into the bottom one, which then becomes buoyant and floats to the top. The squeezing mechanism in these designs would not be able to do enough work to move the air down, or would leave no excess work available to be extracted.

Patents

Proposals for such inoperable machines have become so common that the United States Patent and Trademark Office (USPTO) has made an official policy of refusing to grant patents for perpetual motion machines without a working model. The USPTO Manual of Patent Examining Practice states:

With the exception of cases involving perpetual motion, a model is not ordinarily required by the Office to demonstrate the operability of a device. If operability of a device is questioned, the applicant must establish it to the satisfaction of the examiner, but he or she may choose his or her own way of so doing.

And, further, that: A rejection [of a patent application] on the ground of lack of utility includes the more specific grounds of inoperativeness, involving perpetual motion. A rejection under 35 U.S.C. 101 for lack of utility should not be based on grounds that the invention is frivolous, fraudulent or against public policy.

The filing of a patent application is a clerical task, and the USPTO will not refuse filings for perpetual motion machines; the application will be filed and then most probably rejected by the patent examiner, after he has done a formal examination. Even if a patent is granted, it does not mean that the invention actually works, it just means that the examiner believes that it works, or was unable to figure out why it would not work.

The USPTO maintains a collection of Perpetual Motion Gimmicks.

The United Kingdom Patent Office has a specific practice on perpetual motion; Section 4.05 of the UKPO Manual of Patent Practice states:

Processes or articles alleged to operate in a manner which is clearly contrary to well-established physical laws, such as perpetual motion machines, are regarded as not having industrial application.

Examples of decisions by the UK Patent Office to refuse patent applications for perpetual motion machines include:

Decision BL O/044/06, John Frederick Willmott's application no. 0502841

Decision BL O/150/06, Ezra Shimshi's application no. 0417271

The European Patent Classification (ECLA) has classes including patent applications on perpetual motion systems: ECLA classes "F03B17/04: Alleged perpetuamobilia..." and "F03B17/00B: [... machines or engines] (with closed loop circulation or similar :... Installations wherein the liquid circulates in a closed loop; Alleged perpetuamobilia of this or similar kind...".

Apparent Perpetual Motion Machines

While "perpetual motion" can only exist in isolated systems, and true isolated systems don't exist, there aren't any real "perpetual motion" devices. However there are concepts and technical drafts that propose "perpetual motion", but on closer analysis it's revealed that they actually "consume" some sort of natural resource or latent energy, such as the phase changes of water or other fluids or small natural temperature gradients, or simply can't sustain indefinite operation. In general, extracting large amounts of work using these devices is difficult to impossible.

Resource Consuming

Some examples of such devices include:

- The drinking bird toy functions using small ambient temperature gradients and evaporation.
- A capillarity based water pump functions using small ambient temperature gradients and vapour pressure differences.

- A Crookes radiometer consists of a partial vacuum glass container with a lightweight propeller moved by (light-induced) temperature gradients.
- Any device picking up minimal amounts of energy from the natural electromagnetic radiation around it, such as a solar powered motor.
- Any device powered by changes in air pressure, such as some clocks. The motion leeches energy from moving air which in turn gained its energy from being acted on.
- The Atmos clock uses changes in the vapour pressure of ethyl chloride with temperature to wind the clock spring.
- A device powered by radioactive decay from an isotope with a relatively long half-life; such a device could plausibly operate for hundreds or thousands of years.
- The Oxford Electric Bell and Karpen Pile driven by dry pile batteries.
- Cox's timepiece and the Beverly Clock powered by changes in atmospheric pressure.

Low Friction

- In flywheel energy storage, "modern flywheels can have a zero-load rundown time measurable in years."
- Once spun up, objects in the vacuum of space—stars, black holes, planets, moons, spin-stabilized satellites, etc.—continue spinning almost indefinitely with no further energy input. Tide on Earth is dissipating the gravitational energy of the Moon/ Earth system at an average rate of about 3.75 terawatts.
- In certain quantum-mechanical systems (such as superfluidity and superconductivity), dissipation-free "motion" is possible.

Thought Experiments

In some cases a thought (or "gedanken") experiment appears to suggest that perpetual motion may be possible through accepted and understood physical processes. However, in all cases, a flaw has been found when all of the relevant physics is considered. Examples include:

- Maxwell's Demon: This was originally proposed to show that the Second Law of Thermodynamics applied in the statistical sense only, by postulating a "demon" that could select energetic molecules and extract their energy. Subsequent analysis (and experiment) have shown there is no way to physically implement such a system that does not result in an overall increase in entropy.

- Brownian Ratchet: In this thought experiment, one imagines a paddle wheel connected to a ratchet. Brownian motion would cause surrounding gas molecules to strike the paddles, but the ratchet would only allow it to turn in one direction. A more thorough analysis showed that when a physical ratchet was considered at this molecular scale, Brownian motion would also affect the ratchet and cause it to randomly fail resulting in no net gain. Thus, the device would not violate the Laws of thermodynamics.

Instantaneous Centre of Rotation in Plane Motion and Simple Problems

Instant Centre of Rotation

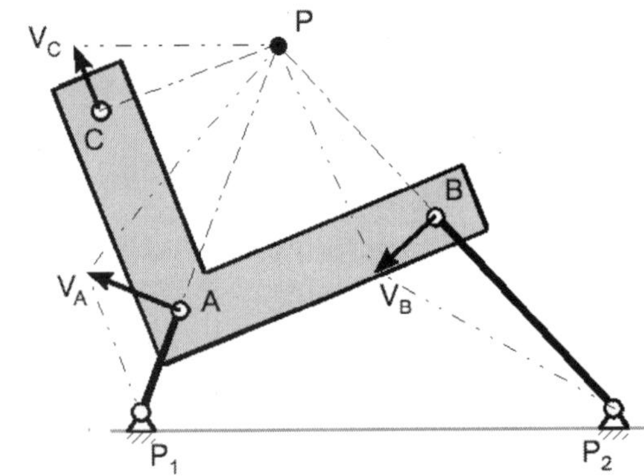

Figure: *Sketch 1: Instant centre of a moving plane*

The instant centre of rotation, also called *instantaneous centre* and *instant centre,* is the point in a body undergoing planar movement that has zero velocity at a particular instant of time. At this instant the velocity vectors of the trajectories of other points in the body generate a circular field around this point which is identical to what is generated by a pure rotation. Planar movement of a body is often described using a plane figure moving in a two dimensional plane. The instant centre is the point in the moving plane around which all other points are rotating at a specific instant of time. The continuous movement of a plane has an instant centre for every value of the time parameter. This generates a curve called the moving centrode. The points in the fixed plane corresponding to these instant centres form the fixed centrode.

Pole of a Planar Displacement

The instant centre can be considered the limiting case of the pole of a planar displacement.

The planar displacement of a body from position 1 to position 2 is defined by the combination of a planar rotation and planar translation. For any planar displacement there is a point in the moving body that is in the same place before and after the displacement. This point is the *pole of the planar displacement*, and the displacement can be viewed as a rotation around this pole.

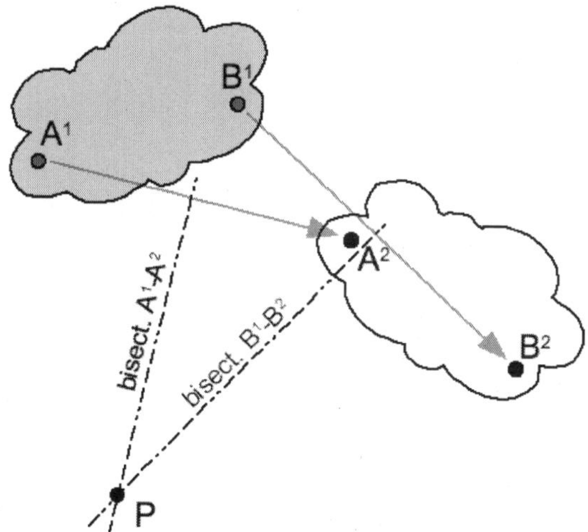

***Figure:** Sketch 2: Pole of a planar displacement*

Construction for the pole of a planar displacement: First, select two points A and B in the moving body and locate the corresponding points in the two positions. Construct the perpendicular bisectors to the two segments A^1A^2 and B^1B^2. The intersection P of these two bisectors is the pole of the planar displacement. Notice that A^1 and A^2 lie on a circle around P. This is true for the corresponding positions of every point in the body.

If the two positions of a body are separated by an instant of time in a planar movement, then the pole of a displacement becomes the instant centre. In this case, the segments constructed between the instantaneous positions of the points A and B become the velocity vectors V_A and V_B. The lines perpendicular to these velocity vectors intersect in the instant centre.

Pure Translation

If the displacement between two positions is a pure translation, then the perpendicular bisectors of the segments A^1B^1 and A^2B^2 form parallel lines. These lines are considered to intersect at a point on the line at infinity, thus the pole of this planar displacement is said to "lie at infinity" in the direction of the perpendicular bisectors.

In the limit, pure translation becomes planar movement with point velocity vectors that are parallel. In this case, the instant centre is said to lie at infinity in the direction perpendicular to the velocity vectors.

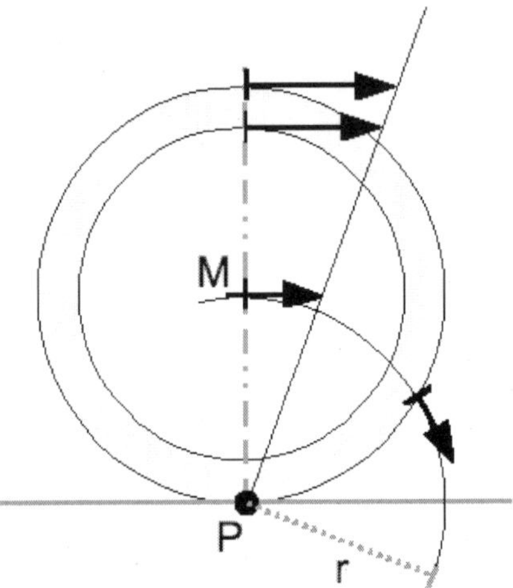

Figure: *Sketch 3: Rolling wheel.*

Instant Centre of a Wheel Rolling without Slipping

Consider the planar movement of a circular wheel rolling without slipping on a linear road. The wheel rotates around its axis M, which translates in a direction parallel to the road. The point of contact P of the wheel with road does not slip, which means the point P has zero velocity with respect to point M. Thus, at the instant the point P on the wheel comes in contact with the road it becomes an instant centre.

The set of points of the moving wheel that become instant centres is the circle itself, which defines the moving centrode. The points in the fixed plane that correspond to these instant centersis the line of the road, which defines the fixed centrode. The velocity vector of a point A in the wheel is perpendicular to the segment AP and is proportional to the length of this segment. In particular, the velocities of points in the wheel are determined by the angular velocity of the wheel in rotation around P. The velocity vectors of a number of points are illustrated in sketch 3.

The further a point in the wheel is from the instant centre P, the proportionally larger its speed. Therefore, the point at the top of the wheel moves in the same direction as the centre M of the wheel, but twice as fast, since it is twice the distance away from P. All points that are a distance equal to the radius of the wheel 'r' from point P move at the same speed as the point M but in different directions. This is shown for a point on the wheel that has the same speed as M but moves in the direction tangent to the circle around P.

Instant Centre of Rotation and Mechanisms

Sketch 1 above shows a four-bar linkage where a number of instant centres of rotation are illustrated. The rigid body noted by the letters BAC is connected with links P_1-A and P_2-B to a base or frame.

The three moving parts of this mechanism (the base is not moving) are: link P_1-A, link P_2-B, and body BAC. For each of these three parts an instant centre of rotation may be determined.

Considering first link P_1-A: all points on this link, including point A, rotate around point P_1. Since P_1 is the only point not moving in the given plane it may be called the instant centre of rotation for this link. Point A, at distance P_1-A from P_1, moves in a circular motion in a direction perpendicular to the link P_1-A, as indicated by vector V_A.

The same applies to link P_2-B: point P_2 is the instant centre of rotation for this link and point B moves in the direction as indicated by vector V_B. For determining the instant centre of rotation of the third element of the linkage, the body BAC, the two points A and B are used because its moving characteristics are known, as derived from the information about the links P_1-A and P_2-B.

The direction of speed of point A is indicated by vector V_A. Its instant centre of rotation must be perpendicular to this vector (as V_A is tangentially located on the circumference of a circle). The only line that fills the requirement is a line colinear with link P_1-A. Somewhere on this line there is a point P, the instant centre of rotation for the body BAC. What applies to point A also applies to point B, therefore this instant centre of rotation P is located on a line perpendicular to vector V_B, a line colinear with link P_2-B. Therefore, the instant centre of rotation P of body BAC is the point where the lines through P_1-A and P_2-B cross. Since this instant centre of rotation P is the centre for all points on the body BAC for any random point, say point C, the speed and direction of movement may be determined: connect P to C. The direction of movement of point C is perpendicular to this connection. The speed is proportional to the distance to point P.

Continuing this approach with the two links P_1-A and P_2-B rotating around their own instant centres of rotation the centrode for instant centre of rotation P may be determined. From this the path of movement for C or any other point on body BAC may be determined.

Examples of Application

In biomechanical research the instant centre of rotation is observed for the functioning of the joints in the upper and lower extremities. For example in analysing the knee, ankle, or shoulder joints. Such knowledge assists in developing artificial joints and prosthesis, such as elbow or finger joints.

Study of the Joints of Horses: *"...velocity vectors determined from the instant centres of rotation indicated that the joint surfaces slide on each other."*.

Studies on turning a vessel moving through water. The braking characteristics of a car may be improved by varying the design of a brake pedal mechanism. Designing the suspension of a bicycle, or of a car. In the case of the coupler link in a four-bar linkage, such as a double wishbone suspension in front view, the perpendiculars to the velocity lie along the links joining the grounded link to the coupler link. This construction is used to establish the kinematic Roll centre of the suspension.

D'Alembert's Principle and its Applications in Plane Motion and Connected Bodies

D'Alembert's principle, also known as the Lagrange–d'Alembert principle, is a statement of the fundamental classical laws of motion. It is named after its discoverer, the Frenchphysicist and mathematician Jean le Rondd' Alembert. The principle states that the sum of the differences between the forces acting on a system of mass particles and the time derivatives of the moment of the system itself along any virtual displacement consistent with the constraints of the system, is zero. Thus, in symbols d'Alembert's principle is written as following,

$$\sum_i (\mathrm{F}_i - m_i \mathrm{a}_i)\cdot\delta \mathrm{r}_i = 0,$$

where

i is an integer used to indicate (via subscript) a variable corresponding to a particular particle in the system,

F_i is the total applied force (excluding constraint forces) on the i-th particle,

m_i is the mass of the i-th particle,

a_i	is the acceleration of the i-th particle,
$m_i a_i$	together as product represents the time derivative of the momentum of the i-th particle, and
δr_i	is the virtual displacement of the i-th particle, consistent with the constraints.

It is the dynamic analogue to the *principle of virtual work for applied forces* in a static system and in fact is more general than Hamilton's principle, avoiding restriction to holonomic systems. A holonomic constraint depends only on the coordinates and time. It does not depend on the velocities. If the negative terms in accelerations are recognized as *inertial forces*, the statement of d'Alembert's principle becomes *The total virtual work of the impressed forces plus the inertial forces vanishes for reversible displacements.*

This above equation is often called d'Alembert's principle, but it was first written in this variational form by Joseph Louis Lagrange.D'Alembert's contribution was to demonstrate that in the totality of a dynamic system the forces of constraint vanish. That is to say that the generalized forces Q_j need not include constraint forces.

General Case with Changing Masses

The general statement of d'Alembert's principle mentions "the time derivatives of the moment of the system". The momentum of the *i*-th mass is the product of its mass and velocity:

$$p_i = m_i v_i$$

and its time derivative is

$$\dot{p}_i = \dot{m}_i v_i + m_i \dot{v}_i \cdot$$

In many applications, the masses are constant and this equation reduces to

$$\dot{p}_i = m_i \dot{v}_i = m_i a_i ,$$

which appears in the formula given above. However, some applications involve changing masses (for example, chains being rolled up or being unrolled) and in those cases both terms $\dot{m}_i v_i$ and $m_i \dot{v}_i$ have to remain present, giving

$$\sum_i (F_i - m_i a_i - \dot{m}_i v_i) \cdot \delta r_i = 0.$$

Derivation for Special Cases

To date nobody has shown that D'Alembert's principle is equivalent to Newton's Second Law. This is true only for some very special cases e.g. rigid body constraints. However, an approximate solution to this problem does exist.

Consider Newton's law for a system of particles, i. The total force on each particle is

$$\mathrm{F}_i^{(T)} = m_i \mathrm{a}_i ,$$

where

$\mathrm{F}_i^{(T)}$	are the total forces acting on the system's particles,
$m_i \mathrm{a}_i$	are the inertial forces that result from the total forces.

Moving the inertial forces to the left gives an expression that can be considered to represent quasi-static equilibrium, but which is really just a small algebraic manipulation of Newton's law:

$$\mathrm{F}_i^{(T)} - m_i \mathrm{a}_i = 0.$$

Considering the virtual work, δW, done by the total and inertial forces together through an arbitrary virtual displacement, $\delta \mathrm{r}_i$, of the system leads to a zero identity, since the forces involved sum to zero for each particle.

$$\delta W = \sum_i \mathrm{F}_i^{(T)} \cdot \delta \mathrm{r}_i - \sum_i m_i \mathrm{a}_i \cdot \delta \mathrm{r}_i = 0$$

The original vector equation could be recovered by recognizing that the work expression must hold for arbitrary displacements. Separating the total forces into applied forces, F_i, and constraint forces, C_i, yields

$$\delta W = \sum_i \mathrm{F}_i \cdot \delta \mathrm{r}_i + \sum_i \mathrm{C}_i \cdot \delta \mathrm{r}_i - \sum_i m_i \mathrm{a}_i \cdot \delta \mathrm{r}_i = 0.$$

If arbitrary virtual displacements are assumed to be in directions that are orthogonal to the constraint forces (which is not usually the case, so this derivation works only for special cases), the constraint forces do no work. Such displacements are said to be *consistent* with the constraints. This leads to the formulation of *d'Alembert's principle*, which states that the difference of applied forces and inertial forces for a dynamic system does no virtual work:.

$$\delta W = \sum_i (\mathrm{F}_i - m_i \mathrm{a}_i) \cdot \delta \mathrm{r}_i = 0.$$

There is also a corresponding principle for static systems called the principle of virtual work for applied forces.

D'Alembert's Principle of Inertial Forces

D'Alembert showed that one can transform an accelerating rigid body into an equivalent static system by adding the so-called "inertial force" and "inertial torque" or moment. The inertial force must act through the centre of mass and the inertial torque can act anywhere. The system can then be analyzed exactly as a static system subjected to this "inertial force and moment" and the external forces. The advantage is that, in the equivalent static system one can take moments about any point (not just the centre of mass). This often leads to simpler calculations because any force (in turn) can be eliminated from the moment equations by choosing the appropriate point about which to apply the moment equation (sum of moments = zero). Even in the course of Fundamentals of Dynamics and Kinematics of machines, this principle helps in analyzing the forces that act on a link of a mechanism when it is in motion. In textbooks of engineering dynamics this is sometimes referred to as *d'Alembert's principle.*

Example for Plane 2D Motion of a Rigid Body

For a planar rigid body, moving in the plane of the body (the x–y plane), and subjected to forces and torques causing rotation only in this plane, the inertial force is

$$F_i = -m\ddot{r}_c$$

where r_c is the position vector of the centre of mass of the body, and m is the mass of the body. The inertial torque (or moment) is

$$T_i = -I\ddot{\theta}$$

where I is the moment of inertia of the body. If, in addition to the external forces and torques acting on the body, the inertia force acting through the centre of mass is added and the inertial torque is added (acting around the centre of mass is as good as anywhere) the system is equivalent to one in static equilibrium. Thus the equations of static equilibrium

$$\begin{aligned} \sum F_x &= 0, \\ \sum F_y &= 0, \\ \sum T &= 0 \end{aligned}$$

hold. The important thing is that $\sum T$ is the sum of torques (or moments, including the inertial moment and the moment of the inertial force) taken about *any* point. The direct application of Newton's laws requires that the angular acceleration equation be applied *only* about the centre of mass.

2

Basic Concepts

Engineering mechanics is the application of mechanics to solve problems involving common engineering elements.

The goal of this Engineering Mechanics course is to expose students to problems in mechanics as applied to plausibly real-world scenarios. Problems of particular types are explored in detail in the hopes that students will gain an inductive understanding of the underlying principles at work; students should then be able to recognize problems of this sort in real-world situations and respond accordingly.

Further, this text aims to support the learning of Engineering Mechanics with theoretical material, general key techniques, and a sufficient number of solved sample problems to satisfy the first objective as outlined above.

Distinction between Branches of Physics

As you see in the diagram mechanics is the first and most fundamental branch of physics, supporting Thermodynamics and Electricity, and including Statics, Dynamics(=Kinematics+kinetics); all which are all highly applicable in engineering. But the most important part of them is statics (study of body at rest) which is not only a base for all others, but also have the highest engineering application.

Physics also involve optics, waves, quantum, and relativity theory, which have no fundamental engineering application yet.

Statics

We describe the motion of bodies using Newton's second law of motion

$$\text{Force} = \text{Mass} * \text{Acceleration} \qquad \text{or} \qquad f = ma$$

Statics deals with the situation where the acceleration (a) is zero - which happens when a body is at rest, or moving with constant velocity. That means that the total force on a body at rest or with constant velocity must be zero. In other words, the sum of the forces on the body must equal zero.

$$\text{Sum of Forces} = 0 \qquad \text{or} \qquad \sum f = 0$$

Engineers typically draw what is called a "free body diagram" to show all forces on a body at rest. These forces are then broken down into vectors consistent with a useful coordinate system and summed in sets (components parallel to each basis vector) which are then set to zero to meet the static constraint of no acceleration being present.

This typically results in sets of equations which can be solved using simple linear algebra techniques or even simple algebra and substitution.

Truss

Forces act along the members, and there are no shear forces or moments. A truss is therefore defined as a system composed entirely of two-force members, which only carry axial loads. The ends of a truss are pinned, so that they don't carry moments. The only reactions at the ends of a truss member are forces. External forces on trusses act only on the end points. Truss problems are solved by the method of sections, where an imaginary cut is made through the member(s) of interest, and global equilibrium of forces and moments are used to determine the forces in the members, or by the method of joints, in which a single joint is isolated and analyzed and the resulting forces (not necessarily with a numerical value) are transferred to adjacent joints, where the process is repeated. The resulting set of equations can then be solved by linear algebra, or substitution.

Chains and Cables

Chains and cables are attached at end points and have a continuous load on them due to their own weight (body forces) or external loads. Let a cable of length L have a load of $w$acting per unit distance between the supports. If the tension in the cable at any point is $T(x)$, then we have, for an infinitesimal length of the cable making an angle $\theta(x)$ with the horizontal,

$$\frac{d}{dx}[T\cos\theta] = 0$$

$$\frac{d}{dx}[T\sin\theta] = w$$

Thus, we have,

$$T\cos\theta = \text{constant} = K$$

$$T\sin\theta = \int w\,dx + C_1$$

or

$$\tan\theta = \frac{1}{K}\left[\int w\,dx + C_1\right]$$

Now

$$\tan\theta = \frac{dy}{dx}.$$

So we have a differential equation for y in terms of x:

$$\frac{dy}{dx} = \frac{1}{K}\left[\int w\,dx + C_1\right]$$

Solving for y,

$$y = \frac{1}{K}\left[\int\left(\int w\,dx\right)dx + C_1 x + C_2\right]$$

If w is a constant, then we have,

$$y = \frac{1}{K}\left[\frac{wx^2}{2} + C_1 x + C_2\right]$$

which is the equation of a parabola.

For a rope where the loading is given in terms of the length of the rope (much more common), i.e., $T \equiv T(s)$ and $\theta \equiv \theta(s)$, we have,

$$\frac{d}{ds}[T\cos\theta] = 0$$

$$\frac{d}{ds}[T\sin\theta] = w(s)$$

where

$$ds = \sqrt{dx^2 + dy^2}$$

Dynamics

While statics deal with the part of mechanics where all objects are stationary, dynamics deals with objects or structures with a non-zero acceleration.

Dynamics may be broken down into kinematics and kinetics. Kinematics deal with displacement, velocities and accelerations without concern for the forces involved. Kinetics deal with the forces and moments involved in making the body move.

System of Forces

Forces

Statics is the study of rigid bodies that are stationary. To be stationary, a rigid body must be in equilibrium. In the language of statics, a stationary rigid body has no *unbalanced forces* acting on it.

Force is a push or a pull that one body exerts on another, including gravitational, electrostatic magnetic and constant influences. Force is a vector quantity, having a magnitude, direction, and point of application.

Strictly speaking, actions of other bodies on a rigid body are known as external forces. If unbalanced, an *external force* will cause motion of the body. *Internal forces* are the forces that hold together parts of a rigid body. Although internal forces can cause deformation of a body, motion is never caused by internal forces.

Forces are frequently represented in terms of unit vectors and force components. A *unit vector* is a vector of unit length directed along a coordinate axis. Unit vectors are used in vector equations to indicate direction without affecting magnitude. In the rectangular coordinate system, there are three unit vectors, i, j, and k.

In two dimensions,

$$F = F_x i + F_y J \quad \text{[Two-dimensional]}$$

Resultant

The *resultant*, or sum, of n two-dimensional forces is equal to the sum of the components.

$$F = i\sum_{i=1}^{n} F_{x,i} + j\sum_{i=1}^{n} F_{y,i} \quad \text{[two-dimensional]}$$

The magnitude of the resultant is

$$R = \sqrt{\left(\sum_{i=1}^{n} F_{x,i}\right)^2 + \left(\sum_{i=1}^{n} F_{y,i}\right)^2}$$

The direction of the resultant is

$$\theta = \tan^{-1}\left(\frac{\sum_{i=1}^{n} F_{y,i}}{\sum_{i=1}^{n} F_{x,i}}\right)$$

Resolution of a Force

The components of a two- or three-dimensional force can be found from its *direction cosines*, the cosines of the true angles made by the force vector with the x-, y-, and z-axes.

$$F_x = F\cos\theta_x$$

$$F_y = F\cos\theta_y$$

$$F_z = F\cos\theta_z$$

Moments

Moment is the name given to the tendency of a force to rotate, turn or twist a rigid body about an actual or assumed pivot point.

However rotation is not required for the moment to exist. When a restrained body is acted upon by a moment, there is no rotation.

Moments have primary dimensions of length $\times$force.

Typical units are foot-pounds, inch-pounds, and Newton-meters.

Moments are vectors. The moment vector, Mo, for a force about a point O is the *cross product* of the force, F, and the vector from point O to the point of application of the force, known as the *position vector*, r. The scalar product $|r|\sin\theta$ is known as the *moment arm*, d.

$$M_O = r \times F$$

$$M_O = |M_O| = |r||F|\sin\theta = d|F| \quad [\theta \leq 180^\circ]$$

Right-hand rule: Place the position and force vectors tail to tail. Close your right hand and position over the pivot point. Rotate the position vector into the force vector and position your hand such that your fingers curl in the same direction as the position vector rotates. Your extended thumb will coincide with the direction of the moment.

The direction cosines of a force can be used to determine the components of the moment about the coordinate axes

$$M_x = M\cos\theta_x$$

$$M_y = M\cos\theta_y$$

$$M_z = M\cos\theta_z$$

Alternately, the following three equations can be used to determine the components of the moment from the component of a force applied at point (x, y, z) referenced to an origin at (0,0,0).

$$M_x = yF_z - zF_y \qquad (1)$$

$$M_y = zF_x - xF_z \tag{2}$$

$$M_z = xF_y - yF_x \tag{3}$$

The resultant moment magnitude can be reconstituted from its components.

$$M = \sqrt{M_x^2 + M_y^2 + M_z^2} \tag{4}$$

Couples

Any point of equal opposite, and parallel forces constitute a *couple.* A couple is equivalent to a single moment vector. Since the two forces are opposite in sign, the x-, y-, z-components of the forces cancel out. Therefore, a body is induced to rotate without translation. A couple can be counteracted only by another couple. A couple can be moved to any location without affecting the equilibrium requirements. (Such a moment is known as a *free moment, moment of a couple, or coupling moment.)*

The equal but opposite forces produce a moment vector M_O of magnitude F_d. The two forces can be replaced by this moment vector that can be moved to any location on a body.

$$M_O = 2rF\sin\theta = F_d$$

The combination of the moved force and the couple is known as *a force-couple system.* Alternately, a force-couple system can be replaced by a single force located a distance $d = M / F$ away.

Systems of Forces

Any collection of forces and moments in three-dimensional space is statically equivalent to a single resultant force vector plus a single resultant moment vector. (Either or both of these resultants can be zero.)

The x-, y-, and z-components of the resultant force are the sums of the x-, y-, and z-components of the individual forces, respectively.

$$R = \sum F_n$$

$$= i\sum_{i=1}^{n} F_{x,i} + j\sum_{i=1}^{n} F_{y,i} + k\sum_{i=1}^{n} F_{z,i} \qquad \text{[Three-dimensional]}$$

The resultant moment vector is more complex. It includes the moments of all system forces around the references axes plus the components of all system moments.

$$M = \sum M_n$$

$$M_x = \sum_i (yF_x - zF_y)i + \sum_i (M\cos\theta_x)i$$

$$M_y = \sum_i (zF_x - xF_z)i + \sum_i (M\cos\theta_y)i$$

$$M_z = \sum_i (xF_y - yF_x)i + \sum_i (M\cos\theta_z)i$$

Equilibrium Requirements

An object is static when it is stationary. To be stationary, all of the forces on the object must be in equilibrium. For an object to be in equilibrium, the resultant force and moment vectors must be both be zero.

R=0

$$R = \sqrt{R_x^2 + R_y^2 + R_z^2} = 0$$

M=0

$$M = \sqrt{M_x^2 + M_y^2 + M_z^2} = 0 \qquad (5)$$

Since the square of any non-zero quantity is positive.

$$R_x = 0$$

$$R_y = 0$$

$$R_z = 0$$

$$M_x = 0$$

$$M_y = 0$$

$$M_z = 0$$

Concurrent Forces

A *concurrent force* system is a category of force systems wherein all of the forces act at the same point.

If the forces on a body are all concurrent forces, then only force equilibrium is necessary to ensure complete equilibrium. In two dimensions,

$$\sum F_x = 0$$

$$\sum F_y = 0$$

In three dimensions,

$$\sum F_x = 0$$
$$\sum F_y = 0$$
$$\sum F_z = 0$$

Problems-solving Approaches

Determinacy

When the equations of equilibrium are independent, a rigid body force system is said to be *statically determinate.* A statically determinate system can be solved for all unknowns, which are usually reactions supporting the body. Examples of determinate beam types are illustrated in figure.

Free-Body Diagrams: A *free-body diagram* is a representation of a body in equilibrium, showing all applied forces, moment, and reactions.

Reactions

The first step in solving most statics problems, after drawing the free-body diagram, is to determine the reaction forces (*i.e., the reactions*) supporting the body.

Procedure for Determining the Forces: The procedure for finding determinate reactions in two-dimensional problems is straightforward. Determinate structures will have either a roller support and pinned support or two roller supports.

Step 1: Establish a convenient set of coordinate axes. (To simplify the analysis, one of the coordinate directions should coincide with the direction of the forces and reactions.)

Step 2: Draw the free-body diagram.

Step 3: Resolve the reaction at the pinned support (if any) into components normal and parallel to the coordinate axes

Step 4: Establish a positive direction of rotation (e.g., clockwise) for purposes of taking the moments.

Step 5: Write the equilibrium equation for moments about the pinned connection.

Step 6: Write the equilibrium equation for the forces in the vertical direction. Usually, this Equation will have two unknown vertical reactions.

Step 7: Substitute the known vertical reaction from step5 into the equilibrium equation from step6. This will determine the second vertical reaction.

Step 8: Write the equilibrium equation for the forces at the horizontal direction.

Step 9: If necessary, combine the vertical and horizontal force components at the pinned connection into a resultant reaction.

Trusses

Statically Determinate Trusses

A *truss or frame* is a set of pin-connected axial *members (i.e., two-force members).*

A *structural cell* consists of all members in a closed loop of members. For the truss to be stable (i.e., to be a *rigid truss*), all of the structural cells must be triangular.

Truss loads are considered to act only in the plane of a truss. Therefore, trusses are analyzed as two-dimensional structures. Forces in truss members hold the various truss parts together and are known as *internal forces.* The internal forces are found by drawing free-body diagrams.

Although free-body diagrams of truss members can be drawn, this is not usually done. Instead, free-body diagrams of the pins. (i.e., the joints) are drawn. A pin in compression will be shown with force arrows pointing toward the pin, away from the member. Similarly a pin in tension will be shown with force arrows pointing away from the pin, toward the member.

With typical bridge trusses supported at the ends and loaded downward at the joints, the upper chords are almost always in compression, and the end panels and lower chords are almost always in tension.

Since truss members are axial members, the forces on the truss joints are concurrent forces. Therefore, only force equilibrium needs to be enforced at each pin: the sum of the forces in each of the coordinate directions equals to zero.

Forces in truss members can sometimes be determined by inspection. One of these cases is a *zero-force member*. A third member framing into a joint already connecting two collinear members carries no internal force unless there is a load applied at that joint. Similarly, both members forming an apex of the truss are zero-force members unless there is a load applied at the apex.

If the left-hand side of Eq. 32.1 is greater than the right hand side (i.e., there are *redundant members*), the truss is statically indeterminate.

If the left-hand side is less than the right-hand side, the truss is unstable and will collapse under certain types of loading.

Methods of Joints

The *method of joints* is one of the methods that can be used to find the internal forces in each truss member. This method is useful when most or all of the truss member forces are to be calculated. Because this method advances from joint to adjacent joint. It is inconvenient when a single isolated member force is to be calculated.

Method of Sections

The *method of sections* is a direct approach to finding forces in any truss member. This method is convenient when only a few truss member forces are unknown.

As with previous method, the first step is to find the support reactions. Then a cut is made through the truss, passing through the unknown member. (Knowing where to cut the truss is the key part of this method. Such knowledge is developed only by repeated practice.) Finally, all three conditions of equilibrium are applied as needed to the remaining truss portion. Since there are three equilibrium equations, the cut cannot pass through more than three members in which the forces are unknown.

Pulleys, Cables, and Friction

Pulleys

A *pulley* (also known as a *sheave*) is used to change the direction of an applied tensile force. A series of pulleys working together (known as a *block* and *tackle*) can also provide *pulley advantage* (i.e., *mechanical advantage*).

If the pulley is attached by a bracket to a fixed location, it is said to be a *fixed pulley*. If the pulley is attached to a load, or if the pulley is free to move, it is known as a *free pulley*.

Most simple problems disregard friction and assume that all ropes (fibre ropes, wire ropes, chains, belts, etc.) are parallel. In such cases, the pulley advantage is equal to the number of ropes coming to and going from the load-carrying pulley. The diameters of the pulleys are not factors in calculating the pulley advantage.

Cables

An *ideal cable* is assumed to be completely flexible, massless, and incapable of elongation; therefore, it acts as an axial tension member between points of concentrated loading. In fact, the term *tension or tensile force* is commonly used in place of member force in dealing with cables.

Coplanar Concurrent Forces

In physics, concurrent forces are a system of forces that act in conjunction at the same point in a body. In lamest terms, the lines of action of each force intersect at a common point. If all the forces pertaining to this system act on the same plane or surface, they are said to be coplanar.

To illustrate forces, which are vector quantities, both the magnitude and direction are required. These quantities can be represented through the use of arrows drawn to scale, where the length of the arrow equals the magnitude and the tip equals the direction. Furthermore, concurrent forces can be summed through a methodical approach. This sum is known as the resultant.

Adding Two Concurrent Forces

Two concurrent forces may be added to find their resultant. This can be done through the use of the parallelogram method. The parallelogram is a graphical approach to understanding a system of forces. Unfortunately, due to its nature, it cannot be derived mathematically and has only been based off of experimental evidence. The resultant in this method is the diagonal of the parallelogram formed by the two forces.

In using the parallelogram method, we understand that concurrent forces may be added by re-arranging the vectors in a tip-to-tail fashion. This is called the triangular method. No matter the order of the forces, the resultant will always be constant as long as this approach is performed appropriately.

A great disadvantage in using the parallelogram and the triangular method is the lack of accuracy. As these methods are graphical and require the subject to physically draw the system with a representative scale, the margin of error is of higher degree. Trigonometric functions are preferred to derive the resultant appropriately.

To solve for two concurrent forces in a system use the law of cosines and the law of sines.

Resultant of Three or more Concurrent Forces

When finding the resultant of three or more concurrent forces, the parallelogram and the triangular method can still be used. However, their application increases consistently in complexity as the number of forces increases. Algebraic calculations through trigonometric functions can still be performed but may become cumbersome depending on the amount of triangles created. Based on the same principles as the triangular method, the polygon method may be preferred. It allows the subject to draw multiple forces to scale in a tip-to-tail fashion. The resultant will always be constant however applied.

The easiest method to solve the resultant for multiple concurrent forces precisely is through rectangular components, finding the sum of the all parallel and perpendicular forces and using using those sums as inputs in the Pythagorean theorem. The angle at which this force is being applied can be found through trigonometric functions.

Components in Space

Resultant

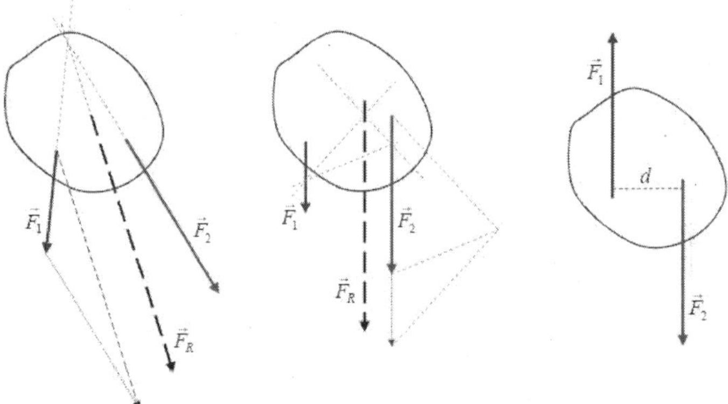

Figure: *Graphical placing of the resultant force*

Resultant forces refers to the reduction of a system of forces acting on a body to a single force and an associated torque. The choice of the point of application of the force determines the associated torque. The term *resultant force* should be understood to refer to both the forces and torques acting on a rigid body, which is why some use the term resultant force-torque.

The resultant force, or resultant force-torque, fully replaces the effects of all forces on the motion of the rigid body they act upon.

Associated Torque

If a point R is selected as the point of application of the resultant force F of a system of *n* forces F_i then the associated torque T is determined from the formulas

$$F = \sum_{i=1}^{n} F_i,$$

and

$$T = \sum_{i=1}^{n} (R_i - R) \times F_i.$$

It is useful to note that the point of application R of the resultant force may be anywhere along the line of action of F without changing the value of the associated torque. To see this add the vector kF to the point of application R in the calculation of the associated torque,

$$T = \sum_{i=1}^{n} (R_i - (R + kF)) \times F_i.$$

The right side of this equation can be separated into the original; formula for T plus the additional term including kF,

$$T = \sum_{i=1}^{n} (R_i - R) \times F_i - \sum_{i=1}^{n} kF \times F_i.$$

Now because F is the sum of the vectors F_i this additional term is zero, that is

$$\sum_{i=1}^{n} kF \times F_i = kF \times (\sum_{i=1}^{n} F_i) = 0,$$

and the value of the associated torque is unchanged.

Torque-free Resultant

It is useful to consider whether there is a point of application R such that the associated torque is zero. This point is defined by property

$$R \times F = \sum_{i=1}^{n} R_i \times F_i,$$

where F is resultant force and F_i form the system of forces.

Notice that this equation for R has a solution only if the sum of the individual torques on the right side yield a vector that is perpendicular to F. Thus, the condition that a system of forces has a torque-free resultant can be written as

$$F \cdot (\sum_{i=1}^{n} R_i \times F_i) = 0.$$

If this condition is not satisfied, then the system of forces includes a pure torque.

The diagram illustrates simple graphical methods for finding the line of application of the resultant force of simple planar systems.

1. Lines of application of the actual forces $\vec{F}_1$ and $\vec{F}_2$ on the leftmost illustration intersect. After vector addition is performed "at the location of $\vec{F}_1$", the net force obtained is translated so that its line of application passes through the common intersection point. With respect to that point all torques are zero, so the torque of the resultant force $\vec{F}_R$ is equal to the sum of the torques of the actual forces.
2. Illustration in the middle of the diagram shows two parallel actual forces. After vector addition "at the location of $\vec{F}_2$", the net force is translated to the appropriate line of application, where it becomes the resultant force $\vec{F}_R$. The procedure is based on decomposition of all forces into components for which the lines of application (pale dotted lines) intersect at one point (the so-called pole, arbitrarily set at the right side of the illustration). Then the arguments from the previous case are applied to the forces and their components to demonstrate the torque relationships.
3. The rightmost illustration shows a couple, two equal but opposite forces for which the amount of the net force is zero, but they produce the net torque $\tau = Fd$ where d is the distance between their lines of application. This is "pure" torque, since there is no resultant force.

Wrench

The forces and torques acting on a rigid body can be assembled into the pair of vectors called a *wrench*. Let *P* be the point of application of the force F and let R be the vector locating this point in a fixed frame. Then the pair of vectors W=(F, R×F) is called a *wrench*. Vectors of this form are known as screws and their mathematics formulation is called screw theory.

The resultant force and torque on a rigid body obtained from a system of forces F_i i=1,...,n, is simply the sum of the individual wrenches W_i, that is

$$W = \sum_{i=1}^{n} W_i = \sum_{i=1}^{n} (F_i, R_i \times F_i).$$

Notice that the case of two equal but opposite forces F and -F acting at points A and B respectively, yields the resultant W=(F-F, A×F - B×

F) = (0, (A-B)×F). This shows that wrenches of the form W=(0, T) can be interpreted as pure torques.

Moment of Forces and its Application

Torque, moment or moment of force, is the tendency of a force to rotate an object about an axis, fulcrum, or pivot. Just as a force is a push or a pull, a torque can be thought of as a twist to an object. Mathematically, torque is defined as the cross product of the lever-arm distance and force, which tends to produce rotation.

Loosely speaking, torque is a measure of the turning force on an object such as a bolt or a flywheel. For example, pushing or pulling the handle of a wrench connected to a nut or bolt produces a torque (turning force) that loosens or tightens the nut or bolt.

The symbol for torque is typically τ, the Greek letter *tau*. When it is called moment, it is commonly denoted *M*.

The magnitude of torque depends on three quantities: the force applied, the length of the *lever arm* connecting the axis to the point of force application, and the angle between the force vector and the lever arm. In symbols:

$$\tau = \mathrm{r} \times \mathrm{F}$$

$$\tau = rF \sin\theta$$

where

τ is the torque vector and τ is the magnitude of the torque,

r is the displacement vector (a vector from the point from which torque is measured to the point where force is applied), and r is the length (or magnitude) of the lever arm vector,

F is the force vector, and F is the magnitude of the force,

× denotes the cross product,

θ is the angle between the force vector and the lever arm vector.

The length of the lever arm is particularly important; choosing this length appropriately lies behind the operation of levers, pulleys, gears, and most other simple machines involving a mechanical advantage.

Terminology

In US mechanical engineering, the term *torque* means 'the resultant moment of a Couple', and (unlike in US physics), the terms *torque* and *moment* are not interchangeable. *Torque* is defined mathematically as the rate of change of angular momentum of an object. The definition of torque states that one or both of the angular velocity or the moment of inertia of an object are changing. And *moment* is the general term used for the tendency of one or more applied forces to rotate an object about an axis, but not necessarily to change the angular momentum of the object (the concept which in physics is called torque).

For example, a rotational force applied to a shaft causing acceleration, such as a drill bit accelerating from rest, the resulting moment is called a *torque.* By contrast, a lateral force on a beam produces a moment (called a bending moment), but since the angular momentum of the beam is not changing, this bending moment is not called a *torque.* Similarly with any force couple on an object that has no change to its angular momentum, such moment is also not called a *torque.*

This article follows the US physics terminology by calling all moments by the term *torque*, whether or not they cause the angular momentum of an object to change.

History

The concept of torque, also called moment or couple, originated with the studies of Archimedes on levers. The rotational analogues of force, mass, and acceleration are torque, moment of inertia and angular acceleration, respectively.

Definition and Relation to Angular Momentum

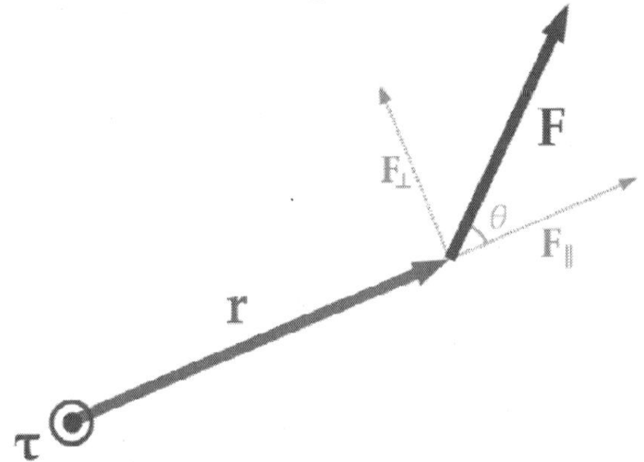

Figure: *A particle is located at position r relative to its axis of rotation. When a force F is applied to the particle, only the perpendicular component* $F_\perp$ *produces a torque. This torque* $\tau = r \times F$ *has magnitude* $\tau = |r||F_\perp| = |r||F|\sin\theta$ *and is directed outward from the page.*

A force applied at a right angle to a lever multiplied by its distance from the lever's fulcrum (the length of the lever arm) is its torque. A force of three newtons applied two metres from the fulcrum, for example, exerts the same torque as a force of one newton applied six metres from the fulcrum. The direction of the torque can be determined by using the right hand grip rule: if the fingers of the right hand are curled from the direction of the lever arm to the direction of the force, then the thumb points in the direction of the torque.

More generally, the torque on a particle (which has the position r in some reference frame) can be defined as the cross product:

$$\tau = \mathrm{r} \times \mathrm{F},$$

where r is the particle's position vector relative to the fulcrum, and F is the force acting on the particle. The magnitude τ of the torque is given by

$$\tau = rF \sin \theta,$$

where r is the distance from the axis of rotation to the particle, F is the magnitude of the force applied, and θ is the angle between the position and force vectors. Alternatively,

$$\tau = rF_{\perp},$$

where $F_{\perp}$ is the amount of force directed perpendicularly to the position of the particle. Any force directed parallel to the particle's position vector does not produce a torque.

It follows from the properties of the cross product that the torque vector is perpendicular to both the position and force vectors. It points along the axis of the rotation that this torque would initiate, starting from rest, and its direction is determined by the right-hand rule.

The unbalanced torque on a body along axis of rotation determines the rate of change of the body's angular momentum,

$$\tau = \frac{\mathrm{dL}}{\mathrm{d}t}$$

where L is the angular momentum vector and t is time. If multiple torques are acting on the body, it is instead the net torque which determines the rate of change of the angular momentum:

$$\tau_1 + \cdots + \tau_n = \tau_{\text{net}} = \frac{\mathrm{dL}}{\mathrm{d}t}.$$

For rotation about a fixed axis,

$$\mathrm{L} - I\omega,$$

where I is the moment of inertia and ω is the angular velocity. It follows that

$$\tau_{net} = \frac{dL}{dt} = \frac{d(I\omega)}{dt} = I\frac{d\omega}{dt} = I\alpha,$$

where α is the angular acceleration of the body, measured in rad/s^2. This equation has the limitation that the torque equation is to be only written about instantaneous axis of rotation or centre of mass for any type of motion - either motion is pure translation, pure rotation or mixed motion. I = Moment of inertia about point about which torque is written (either about instantaneous axis of rotation or centre of mass only). If body is in translatory equilibrium then the torque equation is same about all points in the plane of motion.

Proof of the Equivalence of Definitions

The definition of angular momentum for a single particle is:

$$L = r \times p$$

where "×" indicates the vector cross product, p is the particle's linear momentum, and r is the displacement vector from the origin (the origin is assumed to be a fixed location anywhere in space). The time-derivative of this is:

$$\frac{dL}{dt} = r \times \frac{dp}{dt} + \frac{dr}{dt} \times p.$$

This result can easily be proven by splitting the vectors into components and applying the product rule. Now using the definition of force $F = \frac{dp}{dt}$ (whether or not mass is constant) and the definition of velocity $\frac{dr}{dt} = v$

$$\frac{dL}{dt} = r \times F + v \times p.$$

The cross product of momentum p with its associated velocity v is zero by definition, so the second term vanishes.

By definition, torque $\tau = r \times F$. Therefore torque on a particle is *equal* to the first derivative of its angular momentum with respect to time.

If multiple forces are applied, Newton's second law instead reads $F_{net} = ma$, *and it follows that*

$$\frac{dL}{dt} = r \times F_{net} = \tau_{net}.$$

This is a general proof.

Units

Torque has dimensions of force times distance. Official SI literature suggests using the unit *newton metre* (N·m) or the unit *joule per radian*. The unit *newton metre* is properly denoted N·m or N m. This avoids ambiguity with mN, millinewtons.

The SI unit for energy or work is the joule. It is dimensionally equivalent to a force of one newton acting over a distance of one metre, but it is not used for torque. Energy and torque are entirely different concepts, so the practice of using different unit names (i.e., reserving newton metres for torque and using only joules for energy) helps avoid mistakes and misunderstandings. The dimensional equivalence of these units, of course, is not simply a coincidence: A torque of 1 N·m applied through a full revolution will require an energy of exactly 2π joules. Mathematically,

$$E = \tau\theta$$

where E is the energy, τ is magnitude of the torque, and θ is the angle moved (in radians). This equation motivates the alternate unit name *joules per radian*.

In Imperial units, "pound-force-feet" (lb·ft), "foot-pounds-force", "inch-pounds-force", "ounce-force-inches" (oz·in) are used, and other non-SI units of torque includes "metre-kilograms-force". For all these units, the word "force" is often left out, for example abbreviating "pound-force-foot" to simply "pound-foot" (in this case, it would be implicit that the "pound" is pound-force and not pound-mass). This is an example of the confusion caused by the use of traditional units that may be avoided with SI units because of the careful distinction in SI between force (in newtons) and mass (in kilograms).

Sometimes one may see torque given units that do not dimensionally make sense. For example: gram centimetre. In these units, "gram" should be understood as the force given by the weight of 1 gram at the surface of the earth, i.e., 0.00980665 N. The surface of the earth is understood to have a standard acceleration of gravity (9.80665 m/s^2).

Special Cases and other Facts

Moment arm formula: A very useful special case, often given as the definition of torque in fields other than physics, is as follows:

$$|\tau| = (\text{moment arm})(\text{force}).$$

The construction of the "moment arm" is shown in the figure to the right, along with the vectors r and F mentioned above. The problem with this definition is that it does not give the direction of the torque but only the magnitude, and hence it is difficult to use in three-dimensional cases. If the force is perpendicular to the displacement vector r, the moment arm will be equal to the distance to the centre, and torque will be a maximum for the given force.

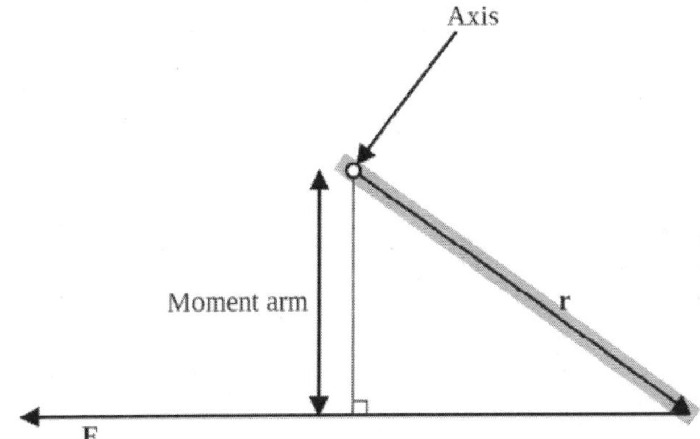

Figure: *Moment arm diagram*

The equation for the magnitude of a torque, arising from a perpendicular force:

$$|\tau| = \text{(distance to centre)(force)}.$$

For example, if a person places a force of 10 N at the terminal end of a spanner (wrench) which is 0.5 m long (or a force of 10 N exactly 0.5 m from the twist point of a spanner of any length), the torque will be 5 N-m – assuming that the person moves the spanner by applying force in the plane of movement of and perpendicular to the spanner.

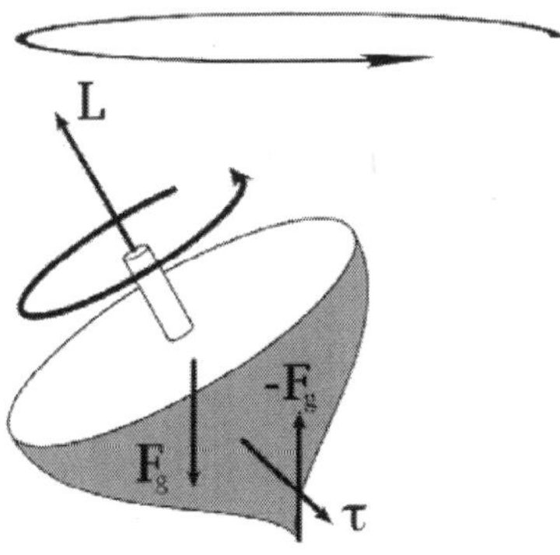

Figure: *The torque caused by the two opposing forces F_g and $-F_g$ causes a change in the angular momentum L in the direction of that torque. This causes the top to precess.*

Static Equilibrium

For an object to be in static equilibrium, not only must the sum of the forces be zero, but also the sum of the torques (moments) about any point. For a two-dimensional situation with horizontal and vertical forces, the sum of the forces requirement is two equations: $\Sigma H = 0$ and $\Sigma V = 0$, and the torque a third equation: $\Sigma\tau = 0$. That is, to solve statically determinate equilibrium problems in two-dimensions, we use three equations.

Net force versus Torque

When the net force on the system is zero, the torque measured from any point in space is the same. For example, the torque on a current-carrying loop in a uniform magnetic field is the same regardless of your point of reference. If the net force F is not zero, and τ_1 is the torque measured from r_1, then the torque measured from r_2 is ... $\tau_2 = \tau_1 + (r_1 - r_2) \times F$

Machine Torque

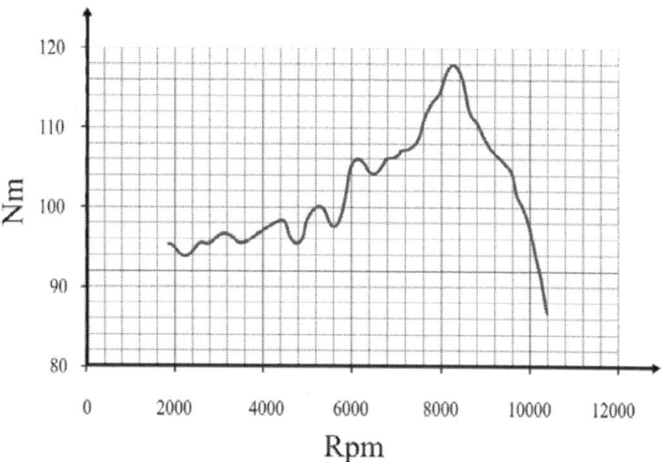

Figure: *Torque curve of a motorcycle ("BMW K 1200 R 2005"). The horizontal axis is the speed (in rpm) that the crankshaft is turning, and the vertical axis is the torque (in Newton metres) that the engine is capable of providing at that speed.*

Torque is part of the basic specification of an engine: the power output of an engine is expressed as its torque multiplied by its rotational speed of the axis. Internal-combustion engines produce useful torque only over a limited range of rotational speeds (typically from around 1,000–6,000 rpm for a small car). The varying torque output over that range can be measured with a dynamometer, and shown as a torque curve.

Steam engines and electric motors tend to produce maximum torque close to zero rpm, with the torque diminishing as rotational speed rises (due to increasing friction and other constraints). Reciprocating steam engines can start heavy loads from zero RPM without a clutch.

Relationship between Torque, Power, and Energy

If a force is allowed to act through a distance, it is doing mechanical work. Similarly, if torque is allowed to act through a rotational distance, it is doing work. Mathematically, for rotation about a fixed axis through the centre of mass,

$$W = \int_{\theta_1}^{\theta_2} \tau \, d\theta,$$

where W is work, τ is torque, and θ_1 and θ_2 represent (respectively) the initial and final angular positions of the body. It follows from the work-energy theorem that W also represents the change in the rotational kinetic energy E_r of the body, given by

$$E_r = \tfrac{1}{2} I \omega^2,$$

where I is the moment of inertia of the body and ω is its angular speed.

Power is the work per unit time, given by

$$P = \tau \cdot \omega,$$

where P is power, τ is torque, ω is the angular velocity, and $\cdot$ represents the scalar product.

Mathematically, the equation may be rearranged to compute torque for a given power output. Note that the power injected by the torque depends only on the instantaneous angular speed – not on whether the angular speed increases, decreases, or remains constant while the torque is being applied (this is equivalent to the linear case where the power injected by a force depends only on the instantaneous speed – not on the resulting acceleration, if any).

In practice, this relationship can be observed in power stations which are connected to a large electrical power grid. In such an arrangement, the generator's angular speed is fixed by the grid's frequency, and the power output of the plant is determined by the torque applied to the generator's axis of rotation.

Consistent units must be used. For metric SI units power is watts, torque is newton metres and angular speed is radians per second (not rpm and not revolutions per second).

Also, the unit newton metre is dimensionally equivalent to the joule, which is the unit of energy. However, in the case of torque, the unit is assigned to a vector, whereas for energy, it is assigned to a scalar.

Conversion to Other Units

A conversion factor may be necessary when using different units of power, torque, or angular speed. For example, if rotational speed (revolutions per time) is used in place of angular speed (radians per time), we multiply by a factor of 2π radians per revolution. In the following formulas, P is power, τ is torque and ω is rotational speed.

$$P = \tau \times 2\pi \times \omega$$

Adding units:

$$P / \text{W} = \tau / (\text{N·m}) \times 2\pi \times \omega / \text{rps}$$

Dividing on the left by 60 seconds per minute gives us the following.

$$P / \text{W} = \frac{\tau / (\text{N·m}) \times 2\pi \times \omega / \text{rpm}}{60}$$

where rotational speed is in revolutions per minute (rpm).

Some people (e.g. American automotive engineers) use horsepower (imperial mechanical) for power, foot-pounds (lbf·ft) for torque and rpm for rotational speed. This results in the formula changing to:

$$P / \text{hp} = \frac{\tau / (\text{lbf·ft}) \times 2\pi \times \omega / \text{rpm}}{33{,}000}.$$

The constant below (in foot pounds per minute) changes with the definition of the horsepower; for example, using metric horsepower, it becomes approximately 32,550.

Use of other units (e.g. BTU per hour for power) would require a different custom conversion factor.

Derivation

For a rotating object, the *linear distance* covered at the circumference of rotation is the product of the radius with the angle covered. That is: linear distance = radius × angular distance. And by definition, linear distance = linear speed × time = radius × angular speed × time.

By the definition of torque: torque = force × radius. We can rearrange this to determine force = torque χ radius. These two values can be substituted into the definition of power:

$$\text{power} = \frac{\text{force} \times \text{linear distance}}{\text{time}} = \frac{\left(\frac{\text{torque}}{r}\right) \times (r \times \text{angular speed} \times t)}{t} = \text{torque} \times \text{angular speed}.$$

The radius r and time t have dropped out of the equation. However, angular speed must be in radians, by the assumed direct relationship between linear speed and angular speed at the beginning of the derivation. If the rotational speed is measured in revolutions per unit of time, the linear speed and distance are increased proportionately by 2π in the above derivation to give:

$$\text{power} = \text{torque} \times 2\pi \times \text{rotational speed}.$$

If torque is in newton metres and rotational speed in revolutions per second, the above equation gives power in newton metres per second or watts. If Imperial units are used, and if torque is in pounds-force feet and rotational speed in revolutions per minute, the above equation gives power in foot pounds-force per minute. The horsepower form of the equation is then derived by applying the conversion factor 33,000 ft ·lbf/min per horsepower:

$$\text{power} = \text{torque} \times 2\pi \times \text{rotational speed} \cdot \frac{\text{ft} \cdot \text{lbf}}{\text{min}} \times \frac{\text{horsepower}}{33{,}000 \cdot \frac{\text{ft} \cdot \text{lbf}}{\text{min}}} \approx \frac{\text{torque} \times \text{RPM}}{5{,}252}$$

because $5252.113122 \approx \dfrac{33{,}000}{2\pi}$.

Principle of Moments

The Principle of Moments, also known as Varignon's theorem (not to be confused with the geometrical theorem of the same name) states that the sum of torques due to several forces applied to *a single* point is equal to the torque due to the sum (resultant) of the forces. Mathematically, this follows from:

$$(\mathrm{r} \times \mathrm{F}_1) + (\mathrm{r} \times \mathrm{F}_2) + \cdots = \mathrm{r} \times (\mathrm{F}_1 + \mathrm{F}_2 + \cdots).$$

Torque Multiplier

A torque multiplier is a tool used to provide a mechanical advantage in applying torque to turn bolts, nuts or other items designed to be actuated by application of torque, such as the actuation of valves, particularly where there are relatively high torque requirements.

Torque multipliers typically employ an epicyclic gear train having one or more stages. Each stage of gearing multiplies the torque applied. In the epicyclic gear systems, torque is applied to the input gear or 'sun' gear. A number of planet gears are arranged around and engaged with this sun gear, and therefore rotate. The outside casing of the multiplier, is also engaged with the planet gear teeth, but is prevented from rotating by means of a reaction arm, causing the planet gears to orbit around the sun gear. The planet gears are held in a 'planetary' carrier which also holds the output drive shaft, therefore as the planet gears orbit around the sun gear, the carrier and so the output shaft rotates. Without the reaction arm to prevent rotation of the outer casing, the output shaft cannot apply torque.

Along with the multiplication of torque, there is a decrease in rotational speed of the output shaft compared to the input shaft. This decrease in speed is inversely proportional to the increase in torque. For example, a torque multiplier with a rating of 3:1 will turn its output shaft with three times the torque, but at one third the speed, of the input shaft. However, due to friction and other inefficiencies in the mechanism, the output torque is slightly lower than the theoretical output.

Couples

In mechanics, a couple is a system of forces with a resultant (a.k.a. net or sum) moment but no resultant force. A better term is force couple or pure moment. Its effect is to create rotation without translation, or more generally without any acceleration of the centre of mass. In rigid body mechanics, force couples are *free vectors*, meaning their effects on a body are independent of the point of application.

The resultant moment of a couple is called a torque. This is not to be confused with the term torque as it is used in physics, where it is merely a synonym of moment. Instead, torque is a *special case* of moment. Torque has special properties that moment does not have, in particular the property of being independent of reference point, as described below.

Simple Couple

Definition- A couple is a pair of forces, equal in magnitude, oppositely directed, and displaced by perpendicular distance or moment.

The simplest kind of couple consists of two equal and opposite forces whose lines of action do not coincide. This is called a "simple couple". The forces have a turning effect or moment called a torque about an axis which is normal to the plane of the forces. The SI unit for the torque of the couple is newton metre.

If the two forces are F and "F, then the magnitude of the torque is given by the following formula:

$$\tau = Fd$$

where

τ is the torque

F is the magnitude of one of the forces

d is the perpendicular distance between the forces, sometimes called the *arm* of the couple

The magnitude of the torque is always equal to *F d*, with the direction of the torque given by the unit vector $\hat{e}$, which is perpendicular to the plane containing the two forces. When d is taken as a vector between the points of action of the forces, then the couple is the cross product of d and F. I.e.,

$$\tau = \mathrm{d} \times \mathrm{F}.$$

Independence of Reference Point

The moment of a force is only defined with respect to a certain point *P* (it is said to be the "moment about *P*"), and in general when *P* is changed, the moment changes. However, the moment (torque) of a *couple* is *independent* of the reference point *P*: Any point will give the same moment. In other words, a torque vector, unlike any other moment vector, is a "free vector".

(This fact is called *Varignon's Second Moment Theorem.*)

The proof of this claim is as follows: Suppose there are a set of force vectors F_1, F_2, etc. that form a couple, with position vectors (about some origin *P*) r_1, r_2, etc., respectively. The moment about *P* is

$$M = r_1 \times F_1 + r_2 \times F_2 + \cdots$$

Now we pick a new reference point *P'* that differs from *P* by the vector r. The new moment is

$$M' = (r_1 + r) \times F_1 + (r_2 + r) \times F_2 + \cdots$$

Now the distributive property of the cross product implies

$$M' = \left(r_1 \times F_1 + r_2 \times F_2 + \cdots\right) + r \times \left(F_1 + F_2 + \cdots\right).$$

However, the definition of a force couple means that

$$F_1 + F_2 + \cdots = 0.$$

Therefore,

$$M' = r_1 \times F_1 + r_2 \times F_2 + \cdots = M$$

This proves that the moment is independent of reference point, which is proof that a couple is a free vector.

Forces and Couples

A force *F* applied to a rigid body at a distance *d* from the centre of mass has the same effect as the same force applied directly to the centre of mass and a couple $C_{\ell} = Fd$. The couple produces an angular acceleration of the rigid body at right angles to the plane of the couple. The force at the centre of mass accelerates the body in the direction of the force without change in orientation. The general theorems are:

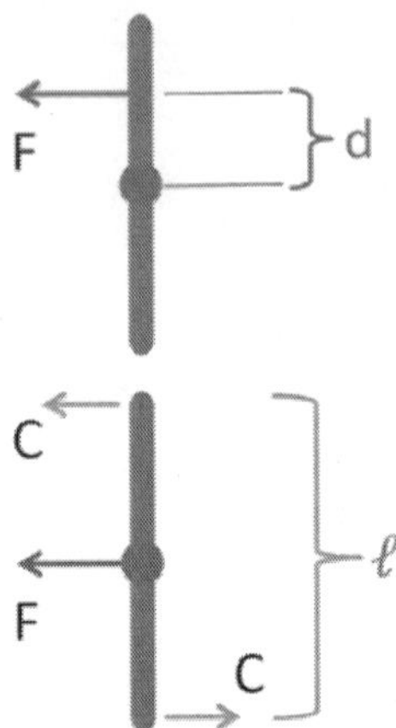

A single force acting at any point O' of a rigid body can be replaced by an equal and parallel force F acting at any given point O and a couple with forces parallel to F whose moment is $M = Fd$, d being the separation of O and O'. Conversely, a couple and a force in the plane of the couple can be replaced by a single force, appropriately located.

Any couple can be replaced by another in the same plane of the same direction and moment, having any desired force or any desired arm.

Applications

Couples are very important in mechanical engineering and the physical sciences. A few examples are:

- The forces exerted by one's hand on a screw-driver
- The forces exerted by the tip of a screw-driver on the head of a screw
- Drag forces acting on a spinning propeller
- Forces on an electric dipole in a uniform electric field.
- The reaction control system on a spacecraft.

In a liquid crystal it is the rotation of an optic axis called the *director* that produces the functionality of these compounds.

At first glance, it may seem that it is optics or electronics which is involved, rather than mechanics. Actually, the changes in optical behaviour, etc. are associated with changes in orientation. In turn, these are produced by couples. Very roughly, it is similar to bending a wire, by applying couples.

Resultant of Force System

A force system is a collection of forces acting at specified locations (may also include couples). Thus the set of forces shown on any free body diagram make up a force system. Force system is simply a term used to describe a group of forces.

When two or more forces or moments are combined, the combination is called a resultant. Any force system can be simplified about a point to a resultant consisting of one force and one couple (either or both of which may be zero). The force is applied to the point and is the vector sum of all of the forces in the system. The couple is the vector sum of all of the couples in the force system plus the vector sum of all of the moments about the point of all of the forces in the force system. One immediate consequence of this definition of resultant is that the resultant about any point of the external force system acting on any system in equilibrium must be zero. The statement that the sum of the external forces and the sum of the external moments acting on a system in equilibrium is zero is often replaced by the statement that the resultant acting on a system in equilibrium is zero. An important attribute of the resultant of a force system is that if you apply the resultant of a force system to a rigid body, the effect on that rigid body is exactly the same as the effect of the original force system. It is in this sense that the resultant is "equivalent" to the original force system. Thus as we study the behaviour of rigid bodies under the action of forces, the resultant of the external force system will be of vital interest. Care must be taken in replacing force systems with their resultant when dealing with deformable bodies as the effects of the resultant acting on a non-rigid system may be different than the effects of the actual force system.

The "equivalence" of effect on a rigid body of a force system and its resultant is the motivation for the term equipollence. Two force systems are said to be equipollent if they have the same resultant about a point. Thus two equipollent force systems have the same effect on a rigid body. Thus the precise way to state what we have learned about the toe is that the upper clamping surface exerts a force system on the toe that is equipollent to a force of 540 lb in the -Y direction acting at a point at the top, middle of the toe. Typically we are not this precise and merely state that the clamping surface is exerting a force of 540 lb in the -Y direction on the top, middle point of the toe. It is important for you to recognize that when you hear such statements, what they mean is that the force system acting on the object is equipollent to the specified force and that the particular distribution or combination of forces involved may be quite complicated.

Equilibrium of System of Forces

A standard definition of static equilibrium is:

A system of particles is in static equilibrium when all the particles of the system are at rest and the total force on each particle is permanently zero. This is a strict definition, and often the term "static equilibrium" is used in a more relaxed manner interchangeably with "mechanical equilibrium", as defined next.

A standard definition of mechanical equilibrium for a particle is:

The necessary and sufficient conditions for a particle to be in mechanical equilibrium is that the net force acting upon the particle is zero. The necessary conditions for mechanical equilibrium for a system of particles are:

(i) the vector sum of all *external forces* is zero and

(ii) the sum of the moments of all *external forces* about any line is zero.

As applied to a rigid body, the necessary and sufficient conditions become:

A rigid body is in mechanical equilibrium when the sum of all forces on all particles of the system is zero, and also the sum of all torques on all particles of the system is zero. A rigid body in mechanical equilibrium is undergoing neither linear nor rotational acceleration; however it could be translating or rotating at a constant velocity. However, this definition is of little use in continuum mechanics, for which the idea of a particle is foreign. In addition, this definition gives no information as to one of the most important and interesting aspects of equilibrium states – their stability. An alternative definition of equilibrium that applies to conservative systems and often proves more useful is:

A system is in mechanical equilibrium if its position in configuration space is a point at which the gradient with respect to the generalized coordinates of the potential energy is zero. Because of the fundamental relationship between force and energy, this definition is equivalent to the first definition. However, the definition involving energy can be readily extended to yield information about the stability of the equilibrium state.

For example, from elementary calculus, we know that a necessary condition for a local minimum *or* a maximum of a differentiable function is a vanishing first derivative (that is, the first derivative is becoming zero). To determine whether a point is a minimum or maximum, one may be able to use the second derivative test. The consequences to the stability of the equilibrium state are as follows:

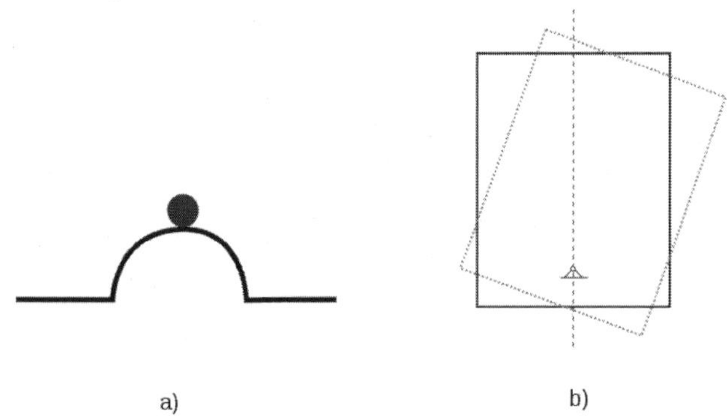

Figure: *Unstable equilibria*

- Second derivative < 0 : The potential energy is at a local maximum, which means that the system is in an unstable equilibrium state. If the system is displaced an arbitrarily small distance from the equilibrium state, the forces of the system cause it to move even farther away.

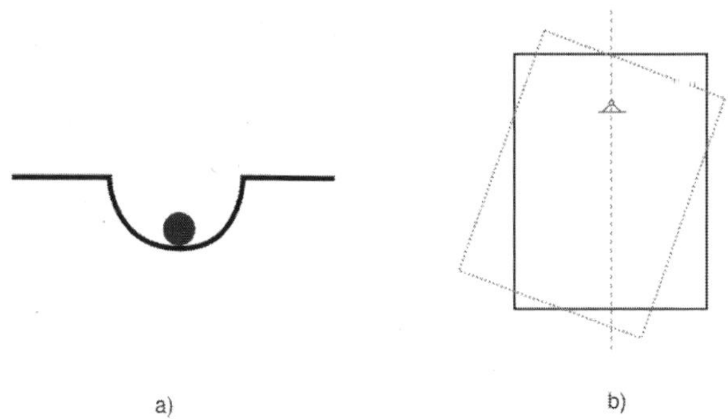

Figure: *Stable equilibria*

- Second derivative > 0 : The potential energy is at a local minimum. This is a stable equilibrium. The response to a small perturbation is forces that tend to restore the equilibrium. If more than one stable equilibrium state is possible for a system, any equilibria whose potential energy is higher than the absolute minimum represent metastable states.

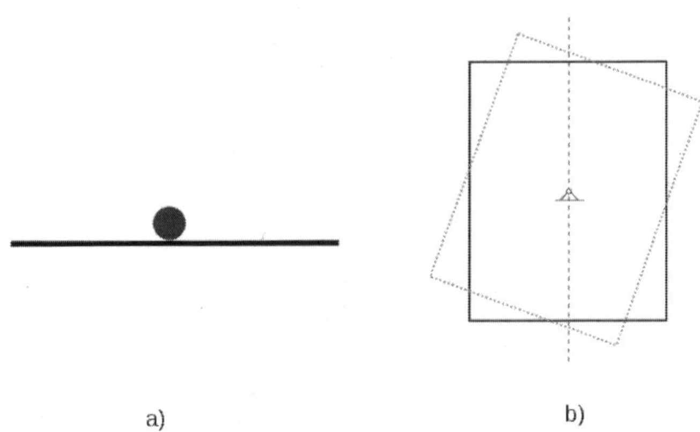

Figure: *Indifferent equilibria*

- Second derivative = 0 or does not exist: The second derivative test fails, and one must typically resort to using the first derivative test. Both of the previous results are still possible, as is a third: this could be a region in which the energy does not vary, in which case the equilibrium is called neutral or indifferent or marginally stable. To lowest order, if the system is displaced a small amount, it will stay in the new state.

In more than one dimension, it is possible to get different results in different directions, for example stability with respect to displacements in the x-direction but instability in the y-direction, a case known as a saddle point. Without further qualification, an equilibrium is stable only if it is stable in all directions.

The special case of mechanical equilibrium of a stationary object is static equilibrium. A paperweight on a desk would be in static equilibrium. The minimal number of static equilibria of homogeneous, convex bodies (when resting under gravity on a horizontal surface) is of special interest. In the planar case, the minimal number is 4, while in three dimensions one can build an object with just one stable and one unstable balance point, this is called Gomboc. A child sliding down a slide at constant speed would be in mechanical equilibrium, but not in static equilibrium.

An example of mechanical equilibrium is a person trying to press a spring. He or she can push it up to a point after which it reaches a state where the force trying to compress it and the resistive force from the spring are equal, so the person cannot further press it. At this state the system will be in mechanical equilibrium. When the pressing force is removed the spring attains its original state.

Free Body Diagrams

A free body diagram, sometimes called a force diagram, is a pictorial device, often a rough working sketch, used by engineers and physicists to analyze the forces and moments acting on a body. The body itself may consist of multiple components, an automobile for example, or just a part of a component, a short section of a beam for example, anything in fact that may be considered to act as a single body, if only for a moment.

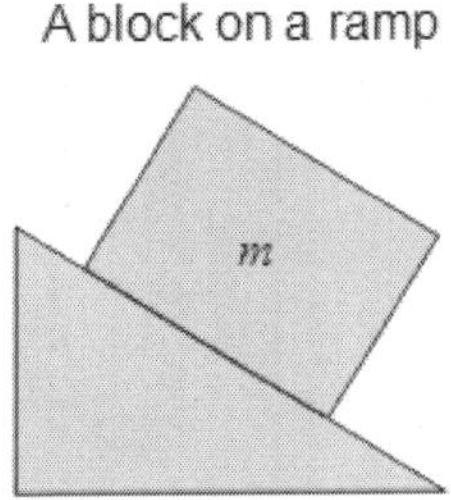

Free body diagram
of just the block

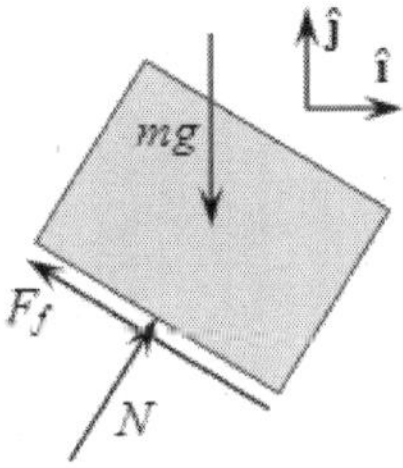

Figure: *Block on a ramp (top) and corresponding free body diagram of just the block (bottom). For equilibrium, the line of action of the three force arrows must intersect at a common point.*

A whole series of such diagrams may be necessary to analyze forces in a complex problem. The free body in a free body diagram is not free of constraints, it is just that the constraints have been replaced by arrows representing the forces and moments they generate.

Purpose

Drawing a free body diagram can help determine the unknown forces on, moments applied to, and equations of motion of, the body and thus help to analyse a problem in statics or dynamics. In analysis of structures, free body diagrams for a component of a structure or, part thereof, are used in determining shear forces and bending moments.

Construction

A free body diagram, an FBD, is not meant to be a scaled drawing. Rather it is a working sketch open to modification as one works through the problem and typically one needs to have seen through the problem before one arrives at a satisfactory diagram. There is an element of art, an inherent flexibility in the whole process. There is no hard and fast algorithm. The iconography of a free body diagram - not only how it is drawn but also how it is interpreted - depends crucially on how a body is modelled.

Modelling the Body

A body may be modelled in three ways:

(i) *a particle.* This model may be used when any turning effects are zero or have zero interest even though the body itself may be extended. The body may be represented by a small symbolic blob and the diagram reduces to a set of concurrent arrows. A force on a particle is a *bound* vector.

(ii) *rigid extended.* Stresses and strains are of no interest but turning effects are. A force arrow should lie along the line of force, but where along the line is irrelevant. A force on an extended rigid body is a *sliding* vector.

(iii) *non-rigid extended.* The *point of application* of a force becomes crucial and has to be indicated on the diagram. A force on a non-rigid body is a *bound* vector. Some engineers use the tail of the arrow to indicate the point of application. Others use the tip.

Example: A Body in Free Fall

The foregoing remarks can be elucidated with the help of an example. Consider a body in free fall in a uniform gravitational field. The body may be

(i) *a particle.* It is enough to show a single vertically downward pointing arrow attached to a blob.

(ii) *rigid extended.* A single arrow suffices to represent the weight W even though gravitational attraction acts on every particle of the body. The arrow must lie along a line through the centre of gravity. Where exactly along the line the arrow should lie is *a matter of convenience.* A popular choice is to locate the arrow at the centre of gravity but strictly speaking there is no need to do so in rigid body analysis.

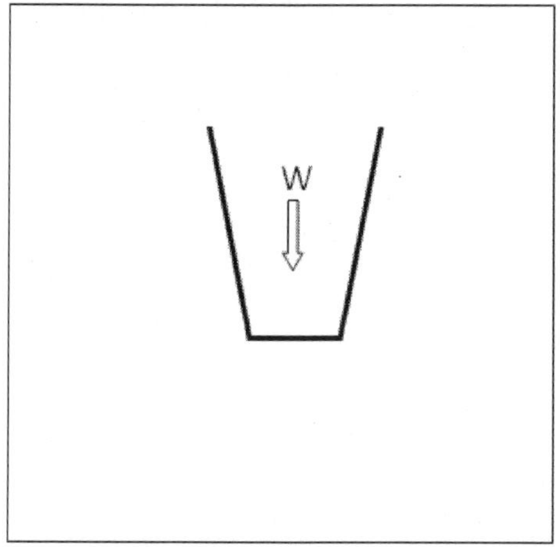

Figure: *An empty rigid bucket in free fall in a uniform gravitational field with the force arrow at the centre of gravity*

(iii) *non-rigid extended.* In non-rigid analysis, it would be a positive error to associate a single point of application with the gravitational force.

What is Included

An FBD represents the body of interest and the external forces on it.

1. *The body:* This is usually sketched in a schematic way depending on the body - particle/extended, rigid/non-rigid - and on what questions are to be answered. Thus if rotation of the body and torque is in consideration, an indication of size and shape of the body is needed. For example The brake dive of a motorcycle cannot be found from a single point, and a sketch with finite dimensions is required.
2. *The external forces:* These are indicated by labelled arrows. In a fully solved problem, a force arrow is capable of indicating
   (i) the direction and the line of action
   (ii) the magnitude
   (iii) the point of application.

Typically, however, a provisional free body sketch is drawn before all these things are known. After all, the whole point of the diagram is to help to determine these things! Thus when a force arrow is originally drawn its length may not be meant to indicate the unknown magnitude. Its line may not correspond to the exact line of action. Even its direction may turn out to wrong. Very often the original direction of the arrow may be directly opposite to the true direction. An engineer may also omit some forces altogether, especially in rigid body analysis where there are paired forces which cancel each other.

The forces acting on the object include friction, gravity, normal force, drag, tension, or a human force due to pushing or pulling. When in a non-inertial reference frame, fictitious forces, such as centrifugal pseudoforce may be appropriate.

A coordinate system is sometimes included, according to convenience. This may make defining the vectors simpler when writing the equations of motion. The x direction might be chosen to point down the ramp in an inclined plane problem, for example. In that case the friction force only has an x component, and the normal force only has a y component. The force of gravity will still have components in both the x and y direction: $mg\sin(\theta)$ in the x and $mg\cos(\theta)$ in the y, where θ is the angle between the ramp and the horizontal.

What is Excluded

Although there is nothing to stop the engineer from having supplementary sketches to help elucidate the problem situation, the free body diagram proper - an FBD - should *not* show

1. bodies other than the free body.
2. forces exerted *by* the free body.
3. internal force exerted by one part of the free body on another part. For example, if an entire truss is being analyzed to find the reaction forces at the supports, the forces between the individual truss members are not included.

Any velocity or acceleration is left out. These may be indicated instead on a companion diagram, called a "Kinetic diagram", "Inertial response diagram", or the equivalent, depending on the author.

A remark is in order concerning the second point 2. above. By Newton's 3rd law if a body A exerts a force on a body B then B exerts an equal and opposite force on A. This equality of two opposite forces which act on two different bodies is often confused for quite a different equality which pertains to a body in equilibrium subject to two equal and opposite forces. A diagram showing the forces exerted *on* and *by* a body is likely to be misconstrued as showing a body in equilibrium and is best avoided.

Example: A Block on an Inclined Plane

A simple free body diagram, shown above, of a block on a ramp illustrates this.

- All external supports and structures have been replaced by the forces they generate. These include:
- *mg*: the product of the mass of the block and the constant of gravitation acceleration: its weight.
- *N*: the normal force of the ramp.
- F_f: the friction force of the ramp.
- The force vectors show direction and point of application and are labelled with their magnitude.
- It contains a coordinate system that can be used when describing the vectors.

Some care is needed in interpreting the diagram. The line of action of the normal force has been shown to be at the midpoint of the base but its true location can only be found if sufficient further data is given. The diagram as it stands would need to be modified were we told that the block is in equilibrium.

There is a potential difficulty also with the arrow representing friction. The engineer who drew this diagram has used the *tip* of the arrow to indicate the point of application of a force.. Now, the tip of the friction arrow is at the highest point of the base. The intention however is not to indicate that the friction acts at that point. The engineer in this instance has assumed a rigid body scenario and that the friction force is a sliding vector and thus the point of application is not relevant. The engineer has tried to indicate that the friction acts all along the whole base by drawing an arrow all along the base but such artistic ploys are a matter of personal choice.

3

Work Energy Principle

In physics, a force is said to do work when it acts on a body so that there is a displacement of the point of application in the direction of the force. Thus a force does work when it results in movement.

The term *work* was introduced in 1826 by the French mathematician Gaspard-GustaveCoriolis as "weight *lifted* through a height", which is based on the use of early steam engines to lift buckets of water out of flooded ore mines. The SI unit of work is the newton-metre or joule (J).

The work done by a constant force of magnitude F on a point that moves a displacement d in the direction of the force is the product,

$$W = Fd.$$

For example, if a force of 10 newton ($F = 10$ N) acts along point that travels 2 metres ($d = 2$ m), then it does the work $W = (10 \text{ N})(2 \text{ m}) = 20 \text{ N m} = 20 \text{ J}$. This is approximately the work done lifting a 1 kg weight from ground to over a person's head against the force of gravity. Notice that the work is doubled either by lifting twice the weight the same distance or by lifting the same weight twice the distance.

Units

The SI unit of work is the joule (J), which is defined as the work expended by a force of one newton through a distance of one metre.

The dimensionally equivalent newton-metre (N·m) is sometimes used as the measuring unit for work, but this can be confused with the unit newton-metre, which is the measurement unit of torque. Usage of N·m is discouraged by the SI authority, since it can lead to confusion as to whether the quantity expressed in newton metres is a torque measurement, or a measurement of energy.

Non-SI units of work include the erg, the foot-pound, the foot-poundal, the kilowatt hour, the litre-atmosphere, and the horsepower-hour. Due to work having the same physical dimension as heat, occasionally measurement units typically reserved for heat or energy content, such as therm, BTU and Calorie, are utilized as a measuring unit.

Work and Energy

Work is closely related to energy. The conservation of energy states that the change in total internal energy of a system equals the added heat, minus the work performed by the system,

$$\delta E = \delta Q - \delta W.$$

Also, from Newton's second law for rigid bodies it can be shown that work on an object is equal to the change in kinetic energy of that object,

$$W = \Delta KE \cdot$$

The work of forces generated by a potential function is known as potential energy and the forces are said to be conservative. Therefore work on an object moving in a conservative force field is equal to *minus* the change of potential energy of the object,

$$W = -\Delta PE.$$

These formulas demonstrate that work is the energy associated with the action of a force, so work subsequently possesses the physical dimensions, and units, of energy. The work/energy principles discussed here are identical to Electric work/energy principles.

Constraint Forces

Constraint forces determine the movement of components in a system, constraining the object within a boundary (in the case of a slope plus gravity, the object is *stuck to* the slope, when attached to a taut string it cannot move in an outwards direction to make the string any 'tauter'). They eliminate all movement in the direction of the constraint, thus constraint forces do not perform work on the system, as the velocity of that object is constrained to be 0 parallel to this force, due to this force.

For example, the centripetal force exerted *inwards* by a string on a ball in uniform circular motion *sideways* constrains the ball to circular motion restricting its movement away from the centre of the circle. This force does zero work because it is perpendicular to the velocity of the ball. Another example is a book on a table. If external forces are applied to the book so that it slides on the table, then the force exerted by the table constrains the book from moving downwards. The force exerted by the table supports the book and is perpendicular to its movement which means that this constraint force does not perform work.

The magnetic force on a charged particle is $F = qv \times B$, where q is the charge, v is the velocity of the particle, and B is the magnetic field. The result of a cross product is always perpendicular to both of the original vectors, so $F \perp v$. The dot product of two perpendicular vectors is always zero, so the work $W = F \cdot v = 0$, and the magnetic force does not do work. It can change the direction of motion but never change the speed.

Mathematical Calculation

For moving objects, the quantity of work/time is calculated as distance/time, or velocity. Thus, at any instant, the rate of the work done by a force (measured in joules/second, or watts) is the scalar product of the force (a vector), and the velocity vector of the point of application. This scalar product of force and velocity is classified as instantaneous power. Just as velocities may be integrated over time to obtain a total distance, by the fundamental theorem of calculus, the total work along a path is similarly the time-integral of instantaneous power applied along the trajectory of the point of application.

Work is the result of a force on a point that moves through a distance. As the point moves, it follows a curve X, with a velocity v, at each instant. The small amount of work δW that occurs over an instant of time δt is calculated as

$$\delta W = F \cdot v \delta t,$$

where the F.v is the power over the instant δt. The sum of these small amounts of work over the trajectory of the point yields the work,

$$W = \int_{t_1}^{t_2} F \cdot v dt = \int_{t_1}^{t_2} F \cdot \tfrac{dx}{dt} dt = \int_C F \cdot dx,$$

where C is the trajectory from $x(t_1)$ to $x(t_2)$. This integral is computed along the trajectory of the particle, and is therefore said to be *path dependent.*

If the force is always directed along this line, and the magnitude of the force is F, then this integral simplifies to

$$W = \int_C F ds$$

where s is distance along the line. If F is constant, in addition to being directed along the line, then the integral simplifies further to

$$W = \int_C F ds = F \int_C ds = Fd$$

where d is the distance travelled by the point along the line.

This calculation can be generalized for a constant force that is not directed along the line, followed by the particle. In this case the dot product $\mathrm{F} \cdot \mathrm{dx} = F\cos\theta \mathrm{dx}$, where θ is the angle between the force vector and the direction of movement, that is

$$W = \int_C \mathrm{F} \cdot d\mathrm{x} = Fd\cos\theta.$$

In the notable case of a force applied to a body always at an angle of 90 degrees from the velocity vector (as when a body moves in a circle under a central force), no work is done at all, since the cosine of 90 degrees is zero. Thus, no work can be performed by gravity on a planet with a circular orbit (this is ideal, as all orbits are slightly elliptical). Also, no work is done on a body moving circularly at a constant speed while constrained by mechanical force, such as moving at constant speed in a friction-less ideal centrifuge.

Calculating the work as "force times straight path segment" would only apply in the most simple of circumstances, as noted above. If force is changing, or if the body is moving along a curved path, possibly rotating and not necessarily rigid, then only the path of the application point of the force is relevant for the work done, and only the component of the force parallel to the application point velocity is doing work (positive work when in the same direction, and negative when in the opposite direction of the velocity). This component of force can be described by the scalar quantity called *scalar tangential component* ($F\cos\theta$, where θ is the angle between the force and the velocity). And then the most general definition of work can be formulated as follows:

Work of a force is the line integral of its scalar tangential component along the path of its application point.

Torque and Rotation

A torque results from equal, and opposite forces, acting on two different points of a rigid body. The sum of these forces cancel, but their effect on the body is the torque T. The work of the torque is calculated as

$$\delta W = \mathrm{T} \cdot \vec{\omega} \delta t,$$

where the T.ω is the power over the instant δt. The sum of these small amounts of work over the trajectory of the rigid body yields the work,

$$W = \int_{t_1}^{t_2} \mathrm{T} \cdot \vec{\omega} dt.$$

This integral is computed along the trajectory of the rigid body with an angular velocity ω that varies with time, and is therefore said to be *path dependent*.

If the angular velocity vector maintains a constant direction, then it takes the form,

$$\vec{\omega} = \dot{\phi}S,$$

where φ is the angle of rotation about the constant unit vector S. In this case, the work of the torque becomes,

$$W = \int_{t_1}^{t_2} T\cdot\vec{\omega}dt = \int_{t_1}^{t_2} T\cdot S\frac{d\phi}{dt}dt = \int_C T\cdot Sd\phi,$$

where C is the trajectory from $\varphi(t_1)$ to $\varphi(t_2)$. This integral depends on the rotational trajectory $\varphi(t)$, and is therefore path-dependent.

If the torque T is aligned with the angular velocity vector so that,

$$T = \tau S,$$

and both the torque and angular velocity are constant, then the work takes the form,

$$W = \int_{t_1}^{t_2} \tau\dot{\phi}dt = \tau(\phi_2 - \phi_1).$$

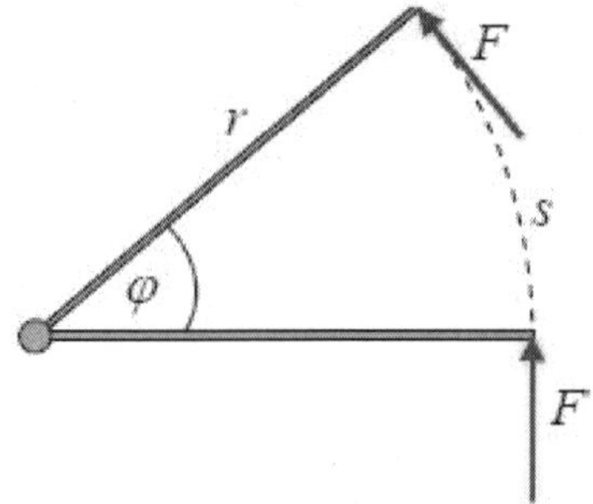

Figure: *A force of constant magnitude and perpendicular to the lever arm*

This result can be understood more simply by considering the torque as arising from a force of constant magnitude F, being applied perpendicularly to a lever arm at a distance r, as shown in the figure. This force will act through the distance along the circular arc $s=r\varphi$, so the work done is

$$W = Fs = Fr\phi.$$

Introduce the torque $\tau=Fr$, to obtain

$$W = Fr\phi = \tau\phi,$$

as presented above. Notice that only the component of torque in the direction of the angular velocity vector contributes to the work.

Work and Potential Energy

The scalar product of a force F and the velocity v of its point of application defines the power input to a system at an instant of time. Integration of this power over the trajectory of the point of application, C=x(t), defines the work input to the system by the force.

Path Dependence

Therefore, the work done by a force F on an object that travels along a curve C is given by the line integral:

$$W = \int_C \mathrm{F}\cdot d\mathrm{x} = \int_{t_1}^{t_2} \mathrm{F}\cdot \mathrm{v}\,dt,$$

where x(t) defines the trajectory C and v is the velocity along this trajectory. In general this integral requires the path along which the velocity is defined, so the evaluation of work is said to be path dependent.

The time derivative of the integral for work yields the instantaneous power,

$$\frac{dW}{dt} = P(t) = \mathrm{F}\cdot\mathrm{v}.$$

Path Independence

If the work for an applied force is independent of the path, then the work done by the force is evaluated at the start and end of the trajectory of the point of application.

This means that there is a function U (x), called a "potential," that can be evaluated at the two points x(t_1) and x(t_2) to obtain the work over any trajectory between these two points. It is tradition to define this function with a negative sign so that positive work is a reduction in the potential, that is

$$W = \int_C \mathrm{F}\cdot d\mathrm{x} = \int_{\mathrm{x}(t_1)}^{\mathrm{x}(t_2)} \mathrm{F}\cdot d\mathrm{x} = U(\mathrm{x}(t_1)) - U(\mathrm{x}(t_2)).$$

The function U(x) is called the potential energy associated with the applied force. Examples of forces that have potential energies are gravity and spring forces.

In this case, the partial derivative of work yields

$$\frac{\partial W}{\partial \mathrm{x}} = -\frac{\partial U}{\partial \mathrm{x}} = -\left(\frac{\partial U}{\partial x}, \frac{\partial U}{\partial y}, \frac{\partial U}{\partial z}\right) = \mathrm{F},$$

and the force F is said to be "derivable from a potential."

Because the potential U defines a force F at every point x in space, the set of forces is called a force field. The power applied to a body by a force field is obtained from the gradient of the work, or potential, in the direction of the velocity V of the body, that is

$$P(t) = -\frac{\partial U}{\partial \mathrm{x}} \cdot \mathrm{v} = \mathrm{F} \cdot \mathrm{v}.$$

Examples of work that can be computed from potential functions are gravity and spring forces.

Work by Gravity

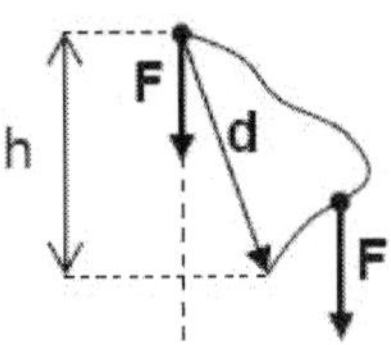

Figure: *Gravity F=mg does work W=mgh along any descending path*

Gravity exerts a constant downward force F=(0, 0, W) on the centre of mass of a body moving near the surface of the earth. The work of gravity on a body moving along a trajectory X(t) = (x(t), y(t), z(t)), such as the track of a roller coaster is calculated using its velocity, v=(v_x, v_y, v_z), to obtain

$$W = \int_{t_1}^{t_2} \mathrm{F} \cdot \mathrm{v} dt = \int_{t_1}^{t_2} W v_z dt = W \Delta z.$$

where the integral of the vertical component of velocity is the vertical distance. Notice that the work of gravity depends only on the vertical movement of the curve X(t).

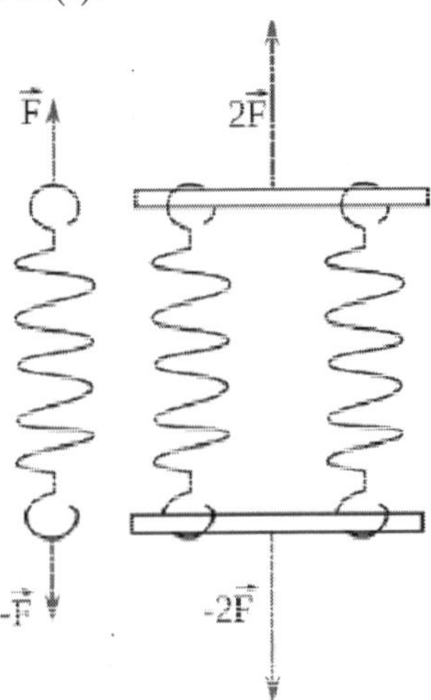

Figure. *Forces in springs assembled in parallel.*

Work by a Spring

A horizontal spring exerts a force F=(kx, 0, 0) that is proportional to its deflection in the x direction. The work of this spring on a body moving along the space curve X(t) = (x(t), y(t), z(t)), is calculated using its velocity, v=(v_x, v_y, v_z), to obtain

$$W = \int_0^t \mathrm{F}\cdot\mathrm{v}dt = \int_0^t kxv_x dt = \frac{1}{2}kx^2.$$

For convenience, consider contact with the spring occurs at t=0, then the integral of the product of the distance x and the x-velocity, xv_x, is $(1/2)x^2$.

Work by a Gas

$$W = \int_a^b PdV$$

Where P is pressure, V is volume, and a and b are initial and final volumes.

Work-energy Principle

The principle of work and kinetic energy (also known as the work-energy principle) states that *the work done by all forces acting on a particle (the work of the resultant force) equals the change in the kinetic energy of the particle.* This definition can be extended to rigid bodies by defining the work of the resultant torque and rotational kinetic energy.

The work W done by the resultant force on a particle equals the change in the particle's kinetic energy E_k,

$$W = \Delta E_k = \tfrac{1}{2}mv_2^2 - \tfrac{1}{2}mv_1^2,$$

where v_1 and v_2 are the speeds of the particle before and after the change and m is its mass.

The derivation of the *work-energy principle* begins with Newton's second law and the resultant force on a particle which includes forces applied to the particle and constraint forces imposed on its movement. Computation of the scalar product of the forces with the velocity of the particle evaluates the instantaneous power added to the system.

Constraints define the direction of movement of the particle by ensuring there is no component of velocity in the direction of the constraint force. This also means the constraint forces do not add to the instantaneous power. The time integral of this scalar equation yields work from the instantaneous power, and kinetic energy from the scalar product of velocity and acceleration. The fact the work-energy principle eliminates the constraint forces underlies Lagrangian mechanics.

This section focusses on the work-energy principle as it applies to particle dynamics. In more general systems work can change the potential energy of a mechanical device, the heat energy in a thermal system, or the electrical energy in an electrical device. Work transfers energy from one place to another or one form to another.

Derivation for a Particle Moving along a Straight Line

In the case the resultant force F is constant in both magnitude and direction, and parallel to the velocity of the particle, the particle is moving with constant acceleration a along a straight line. The relation between the net force and the acceleration is given by the equation $F = ma$ (Newton's second law), and the particle displacement d can be expressed by the equation

$$d = \frac{v_2^2 - v_1^2}{2a}$$

which follows from $v_2^2 = v_1^2 + 2ad$.

The work of the net force is calculated as the product of its magnitude and the particle displacement. Substituting the above equations, one obtains:

$$W = Fd = mad = ma\left(\frac{v_2^2 - v_1^2}{2a}\right) = \frac{mv_2^2}{2} - \frac{mv_1^2}{2} = \Delta E_k$$

In the general case of rectilinear motion, when the net force F is not constant in magnitude, but is constant in direction, and parallel to the velocity of the particle, the work must be integrated along the path of the particle:

$$W = \int_{t_1}^{t_2} \mathrm{F}\cdot\mathrm{v}dt = \int_{t_1}^{t_2} F\,vdt = \int_{t_1}^{t_2} mavdt = m\int_{t_1}^{t_2} v\frac{dv}{dt}dt = m\int_{v_1}^{v_2} vdv = \tfrac{1}{2}m(v_2^2 - v_1^2).$$

General Derivation of the Work-energy Theorem for a Particle

For any net force acting on a particle moving along any curvilinear path, it can be demonstrated that its work equals the change in the kinetic energy of the particle by a simple derivation analogous to the equation above. Some authors call this result *work-energy principle*, but it is more widely known as the work-energy theorem:

$$W = \int_{t_1}^{t_2} \mathrm{F}\cdot\mathrm{v}dt = m\int_{t_1}^{t_2} \mathrm{a}\cdot\mathrm{v}dt = \frac{m}{2}\int_{t_1}^{t_2} \frac{dv^2}{dt}dt = \frac{m}{2}\int_{v_1^2}^{v_2^2} dv^2 = \frac{mv_2^2}{2} - \frac{mv_1^2}{2} = \Delta E_k$$

The identity $a{\cdot}v = \frac{1}{2}\frac{dv^2}{dt}$ requires some algebra. From the identity $v^2 = v{\cdot}v$ and definition $a = \frac{dv}{dt}$ it follows

$$\frac{dv^2}{dt} = \frac{d(v{\cdot}v)}{dt} = \frac{dv}{dt}{\cdot}v + v{\cdot}\frac{dv}{dt} = 2\frac{dv}{dt}{\cdot}v = 2a{\cdot}v\,.$$

The remaining part of the above derivation is just simple calculus, same as in the preceding rectilinear case.

Derivation for a Particle in Constrained Movement

In particle dynamics, a formula equating work applied to a system to its change in kinetic energy is obtained as a first integral of Newton's second law of motion. It is useful to notice that the resultant force used in Newton's laws can be separated into forces that are applied to the particle and forces imposed by constraints on the movement of the particle. Remarkably, the work of a constraint force is zero, therefore only the work of the applied forces need be considered in the work-energy principle.

To see this, consider a particle *P* that follows the trajectory X(t) with a force F acting on it. Isolate the particle from its environment to expose constraint forces *R*, then Newton's Law takes the form

$$F + R = m\ddot{X},$$

where *m* is the mass of the particle.

Vector Formulation

Note that n dots above a vector indicates its nth time derivative. The scalar product of each side of Newton's law with the velocity vector yields

$$F{\cdot}\dot{X} = m\ddot{X}{\cdot}\dot{X},$$

because the constraint forces are perpendicular to the particle velocity. Integrate this equation along its trajectory from the point $X(t_1)$ to the point $X(t_2)$ to obtain

$$\int_{t_1}^{t_2} F{\cdot}\dot{X}\,dt = m\int_{t_1}^{t_2} \ddot{X}{\cdot}\dot{X}\,dt.$$

The left side of this equation is the work of the applied force as it acts on the particle along the trajectory from time t_1 to time t_2. This can also be written as

$$W = \int_{t_1}^{t_2} \mathrm{F} \cdot \dot{\mathrm{X}}\, dt = \int_{\mathrm{X}(t_1)}^{\mathrm{X}(t_2)} \mathrm{F} \cdot d\mathrm{X}.$$

This integral is computed along the trajectory X(t) of the particle and is therefore path dependent.

The right side of the first integral of Newton's equations can be simplified using the following identity

$\frac{1}{2}\frac{d}{dt}(\dot{\mathrm{X}} \cdot \dot{\mathrm{X}}) = \ddot{\mathrm{X}} \cdot \dot{\mathrm{X}},$. Now it is integrated explicitly to obtain the change in kinetic energy,

$$\Delta K = m\int_{t_1}^{t_2} \ddot{\mathrm{X}} \cdot \dot{\mathrm{X}}\, dt = \frac{m}{2}\int_{t_1}^{t_2} \frac{d}{dt}(\dot{\mathrm{X}} \cdot \dot{\mathrm{X}}) dt = \frac{m}{2}\dot{\mathrm{X}} \cdot \dot{\mathrm{X}}(t_2) - \frac{m}{2}\dot{\mathrm{X}} \cdot \dot{\mathrm{X}}(t_1) = \frac{1}{2}m\Delta \mathrm{v}^2,$$

where the kinetic energy of the particle is defined by the scalar quantity,

$$K = \frac{m}{2}\dot{\mathrm{X}} \cdot \dot{\mathrm{X}} = \frac{1}{2}m\mathrm{v}^2$$

Tangential and Normal Components

It is useful to resolve the velocity and acceleration vectors into tangential and normal components along the trajectory X(t), such that

$$\dot{\mathrm{X}} = v\mathrm{T}, \quad \text{and} \quad \ddot{\mathrm{X}} = \dot{v}\mathrm{T} + v^2\kappa\mathrm{N}.$$

where

$$v = |\dot{\mathrm{X}}| = \sqrt{\dot{\mathrm{X}} \cdot \dot{\mathrm{X}}}.$$

Then, the scalar product of velocity with acceleration in Newton's second law takes the form

$$\Delta K = m\int_{t_1}^{t_2} \dot{v}v dt = \frac{m}{2}\int_{t_1}^{t_2} \frac{d}{dt}v^2 dt = \frac{m}{2}v^2(t_2) - \frac{m}{2}v^2(t_1), .$$

where the kinetic energy of the particle is defined by the scalar quantity,

$$K = \frac{m}{2}v^2 = \frac{m}{2}\dot{\mathrm{X}} \cdot \dot{\mathrm{X}}.$$

The result is the work-energy principle for particle dynamics,

$$W = \Delta K.$$

This derivation can be generalized to arbitrary rigid body systems.

Moving in a Straight Line (skid to a stop)

Consider the case of a vehicle moving along a straight horizontal trajectory under the action of a driving force and gravity that sum to

F. The constraint forces between the vehicle and the road define R, and we have

$$\mathrm{F} + \mathrm{R} = m\ddot{\mathrm{X}}.$$

For convenience let the trajectory be along the X-axis, so X=(d,0) and the velocity is V=(v, 0), then R.V=0, and F.V=F_xv, where F_x is the component of F along the X-axis, so

$$F_x v = m\dot{v}v.$$

Integration of both sides yields

$$\int_{t_1}^{t_2} F_x v dt = \frac{m}{2}v^2(t_2) - \frac{m}{2}v^2(t_1).$$

If F_x is constant along the trajectory, then the integral of velocity is distance, so

$$F_x(d(t_2) - d(t_1)) = \frac{m}{2}v^2(t_2) - \frac{m}{2}v^2(t_1).$$

As an example consider a car skidding to a stop, where *k* is the coefficient of friction and *W* is the weight of the car. Then the force along the trajectory is F_x =-kW. The velocity *v* of the car can be determined from the length *s* of the skid using the work-energy principle,

$$kWs = \frac{W}{2g}v^2, \quad \text{or} \quad v = \sqrt{2ksg}.$$

Notice that this formula uses the fact that the mass of the vehicle is $m=W/g$.

Coasting Down a Mountain Road (gravity racing)

Consider the case of a vehicle that starts at rest and coasts down a mountain road, the work-energy principle helps compute the minimum distance that the vehicle travels to reach a velocity *V*, of say 60 mph (88 fps). Rolling resistance and air drag will slow the vehicle down so the actual distance will be less than if these forces are neglected.

Let the trajectory of the vehicle following the road be X(t) which is a curve in three-dimensional space. The force acting on the vehicle that pushes it down the road is the constant force of gravity F=(0,0,*W*), while the force of the road on the vehicle is the constraint force R. Newton's second law yields,

$$\mathrm{F} + \mathrm{R} = m\ddot{\mathrm{X}}.$$

The scalar product of this equation with the velocity, V=(v_x, v_y,v_z), yields

$$Wv_z = m\dot{V}V,$$

where V is the magnitude of V. The constraint forces between the vehicle and the road cancel from this equation because R.V=0, which means they do no work. Integrate both sides to obtain

$$\int_{t_1}^{t_2} Wv_z dt = \frac{m}{2}V^2(t_2) - \frac{m}{2}V^2(t_1).$$

The weight force W is constant along the trajectory and the integral of the vertical velocity is the vertical distance, therefore,

$$W\Delta z = \frac{m}{2}V^2.$$

Recall that $V(t_1)=0$. Notice that this result does not depend on the shape of the road followed by the vehicle.

In order to determine the distance along the road assume the downgrade is 6%, which is a steep road. This means the altitude decreases 6 feet for every 100 feet travelled—for angles this small the sin and tan functions are approximately equal. Therefore, the distance s in feet down a 6% grade to reach the velocity V is at least

$$s = \frac{\Delta z}{0.06} = 8.3\frac{V^2}{g}, \quad \text{or} \quad s = 8.3\frac{88^2}{32.2} \approx 2000\text{ft}.$$

This formula uses the fact that the weight of the vehicle is $W=mg$.

Work of Forces Acting on a Rigid Body

The work of forces acting at various points on a single rigid body can be calculated from the work of a resultant force and torque. To see this, let the forces F_1, F_2... F_n act on the points X_1, X_2... X_n in a rigid body.

The trajectories of X_i, i=1,...,n are defined by the movement of the rigid body. This movement is given by the set of rotations [A(t)] and the trajectory d(t) of a reference point in the body. Let the coordinates x_ii=1,...,n define these points in the moving rigid body's reference frame M, so that the trajectories traced in the fixed frame F are given by

$$X_i(t) = [A(t)]x_i + d(t) \quad i = 1,\ldots,n.$$

The velocity of the points X_i along their trajectories are

$$V_i = \vec{\omega} \times (X_i - d) + \dot{d},$$

where ω is the angular velocity vector obtained from the skew symmetric matrix

$$[\Omega] = \dot{A}A^T,$$

known as the angular velocity matrix.

The small amount of work by the forces over the small displacements δr_i can be determined by approximating the displacement by $\delta r = v\delta t$ so

$$\delta W = F_1 \cdot V_1 \delta t + F_2 \cdot V_2 \delta t + \ldots + F_n \cdot V_n \delta t$$

or

$$\delta W = \sum_{i=1}^{n} F_i \cdot (\vec{\omega} \times (X_i - d) + \dot{d})\delta t.$$

This formula can be rewritten to obtain

$$\delta W = (\sum_{i=1}^{n} F_i) \cdot \dot{d}\,\delta t + (\sum_{i=1}^{n} (X_i - d) \times F_i) \cdot \vec{\omega}\delta t = (F \cdot \dot{d} + T \cdot \vec{\omega})\delta t,$$

where F and T are the resultant force and torque applied at the reference point d of the moving frame *M* in the rigid body.

Kinetics of Rigid Body Rotation

Rigid-body dynamics studies the movement of systems of interconnected bodies under the action of external forces. The assumption that the bodies are rigid, which means that they do not deform under the action of applied forces, simplifies the analysis by reducing the parameters that describe the configuration of the system to the translation and rotation of reference frames attached to each body.

The dynamics of a rigid body system is defined by its equations of motion, which are derived using either Newtonslaws of motion or Lagrangian mechanics. The solution of these equations of motion defines how the configuration of the system of rigid bodies changes as a function of time. The formulation and solution of rigid body dynamics is an important tool in the computer simulation of mechanical systems.

Planar Rigid Body Dynamics

If a rigid system of particles moves such that the trajectory of every particle is parallel to a fixed plane, the system is said to be constrained to planar movement. In this case, Newton's laws for a rigid system of N particles, P_i, i=1,...,N, simplify because there is no movement in the *k* direction. Determine the resultant force and torque at a reference point R, to obtain

$$F = \sum_{i=1}^{N} m_i A_i, \quad T = \sum_{i=1}^{N} (r_i - R) \times (m_i A_i),$$

where r_i denotes the planar trajectory of each particle.

The kinematics of a rigid body yields the formula for the acceleration of the particle P_i in terms of the position R and acceleration A of the reference particle as well as the angular velocity vector ω and angular acceleration vector α of the rigid system of particles as,

$$A_i = \alpha \times (r_i - R) + \omega \times \omega \times (r_i - R) + A.$$

For systems that are constrained to planar movement, the angular velocity and angular acceleration vectors are directed along k perpendicular to the plane of movement, which simplifies this acceleration equation. In this case, the acceleration vectors can be simplified by introducing the unit vectors e_i from the reference point R to a point r_i and the unit vectors $t_i = k x e_i$, so

$$A_i = \alpha(\Delta r_i t_i) - \omega^2(\Delta r_i e_i) + A.$$

This yields the resultant force on the system as

$$F = \alpha \sum_{i=1}^{N} m_i(\Delta r_i t_i) - \omega^2 \sum_{i=1}^{N} m_i(\Delta r_i e_i) + (\sum_{i=1}^{N} m_i)A,$$

and torque as

$$T = \sum_{i=1}^{N} (m_i \Delta r_i e_i) \times (\alpha(\Delta r_i t_i) - \omega^2(\Delta r_i e_i) + A) = (\sum_{i=1}^{N} m_i \Delta r_i^2)\alpha\vec{k} + (\sum_{i=1}^{N} m_i \Delta r_i e_i) \times A,$$

where $e_i x e_i = 0$, and $e_i x t_i = k$ is the unit vector perpendicular to the plane for all of the particles P_i.

Use the centre of massC as the reference point, so these equations for Newton's laws simplify to become

$$F = MA, \quad T = I_C \alpha \vec{k},$$

where M is the total mass and I_C is the moment of inertia about an axis perpendicular to the movement of the rigid system and through the centre of mass.

Rigid Body in Three Dimensions

Orientation or Attitude Descriptions: Several methods to describe orientations of a rigid body in three dimensions have been developed. They are summarized in the following sections.

Euler Angles

The first attempt to represent an orientation was owed to Leonhard Euler. He imagined three reference frames that could rotate one around the other, and realized that by starting with a fixed reference frame and performing three rotations, he could get any other reference frame in the space (using two rotations to fix the vertical axis and other to fix the other two axes).

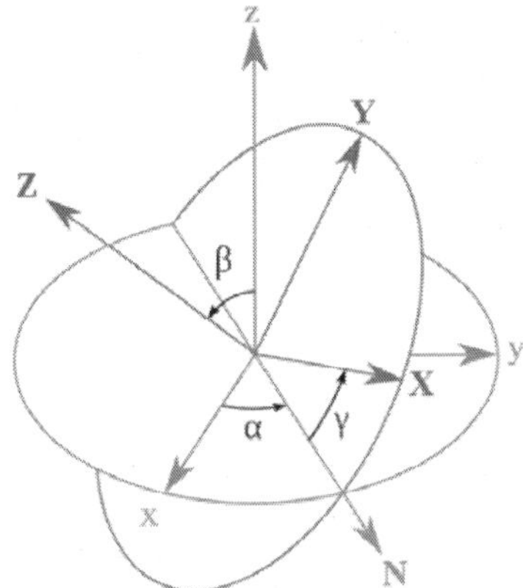

Figure: *Euler angles, one of the possible ways to describe an orientation*

The values of these three rotations are called Euler angles.

Tait-Bryan Angles

These are three angles, also known as yaw, pitch and roll, Navigation angles and Cardan angles. Mathematically they constitute a set of six possibilities inside the twelve possible sets of Euler angles, the ordering being the one best used for describing the orientation of a vehicle such as an airplane. In aerospace engineering they are usually referred to as Euler angles.

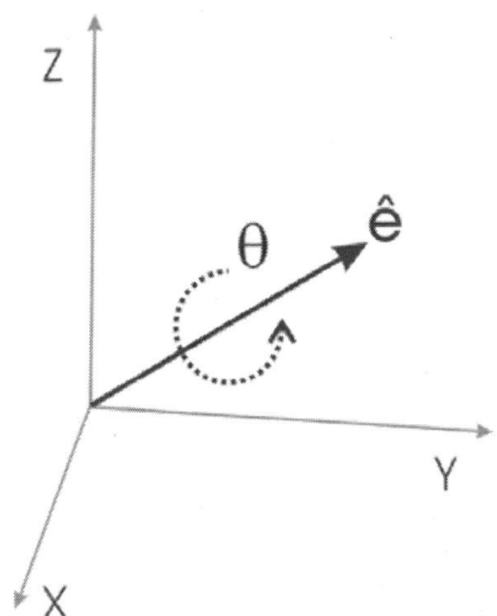

Figure: *A rotation represented by an Euler axis and angle.*

Orientation Vector

Euler also realized that the composition of two rotations is equivalent to a single rotation about a different fixed axis (Euler's rotation theorem). Therefore the composition of the former three angles has to be equal to only one rotation, whose axis was complicated to calculate until matrices were developed. Based on this fact he introduced a vectorial way to describe any rotation, with a vector on the rotation axis and module equal to the value of the angle. Therefore any orientation can be represented by a rotation vector (also called Euler vector) that leads to it from the reference frame. When used to represent an orientation, the rotation vector is commonly called orientation vector, or attitude vector. A similar method, called axis-angle representation, describes a rotation or orientation using a unit vector aligned with the rotation axis, and a separate value to indicate the angle.

Orientation Matrix

With the introduction of matrices the Euler theorems were rewritten. The rotations were described by orthogonal matrices referred to as rotation matrices or direction cosine matrices. When used to represent an orientation, a rotation matrix is commonly called orientation matrix, or attitude matrix.

The above mentioned Euler vector is the eigenvector of a rotation matrix (a rotation matrix has a unique real eigenvalue). The product of two rotation matrices is the composition of rotations. Therefore, as before, the orientation can be given as the rotation from the initial frame to achieve the frame that we want to describe.

The configuration space of a non-symmetrical object in n-*dimensional space is* $SO(n) \times R^n$. *Orientation may be visualized by attaching a basis of tangent vectors to an object. The direction in which each vector points determines its orientation.*

Orientation Quaternion

Another way to describe rotations is using rotation quaternions, also called versors. They are equivalent to rotation matrices and rotation vectors. With respect to rotation vectors, they can be more easily converted to and from matrices. When used to represent orientations, rotation quaternions are typically called orientation quaternions or attitude quaternions.

Newton's Second Law in Three Dimensions

To consider rigid body dynamics in three dimensional space, Newton's second law must be extended to define the relationship between the movement of a rigid body and the system of forces and torques that act on it.

Newton's formulated his second law for a particle as, "The change of motion of an object is proportional to the force impressed and is made in the direction of the straight line in which the force is impressed." Because Newton generally referred to mass times velocity as the "motion" of a particle, the phrase "change of motion" refers to the mass times acceleration of the particle, and so this law is usually written as

$$\mathrm{F} = m\mathrm{a},$$

where F is understood to be the only external force acting on the particle, m is the mass of the particle, and a is its acceleration vector. The extension of Newton's second law to rigid bodies is achieved by considering a rigid system of particles.

Rigid System of Particles

If a system of N particles, P_i, i=1,...,N, are assembled into a rigid body, then Newton's second law can be applied to each of the particles in the body. If F_i is the external force applied to particle P_i with mass m_i, then

$$\mathrm{F}_i + \sum_{j=1}^{N} \mathrm{F}_{ij} = m_i \mathrm{a}_i, \quad i = 1, \ldots, N,$$

where F_{ij} is the internal force of particle P_j acting on particle P_i that maintains the constant distance between these particles.

An important simplification to these force equations is obtained by introducing the resultant force and torque that acts on the rigid system. This resultant force and torque is obtained by choosing one of the particles in the system as a reference point, R, where each of the external forces are applied with the addition of an associated torque. The resultant force F and torque T are given by the formulas,

$$\mathrm{F} = \sum_{i=1}^{N} \mathrm{F}_i, \quad \mathrm{T} = \sum_{i=1}^{N} (\mathrm{R}_i - \mathrm{R}) \times \mathrm{F}_i,$$

where R_i is the vector that defines the position of particle to P_i.

Newton's second law for a particle combines with these formulas for the resultant force and torque to yield,

$$\mathrm{F} = \sum_{i=1}^{N} m_i \mathrm{a}_i, \quad \mathrm{T} = \sum_{i=1}^{N} (\mathrm{R}_i - \mathrm{R}) \times (m_i \mathrm{a}_i),$$

where the internal forces F_{ij} cancel in pairs. The kinematics of a rigid body yields the formula for the acceleration of the particle P_i in terms of the position R and acceleration a of the reference particle as well as the angular velocity vector ω and angular acceleration vector α of the rigid system of particles as,

$$a_i = \alpha \times (R_i - R) + \omega \times \omega \times (R_i - R) + a.$$

Mass Properties

The mass properties of the rigid body are represented by its centre of mass and inertia matrix. Choose the reference point R so that it satisfies the condition

$$\sum_{i=1}^{N} m_i (R_i - R) = 0,$$

then it is known as the centre of mass of the system. The inertia matrix $[I_R]$ of the system relative to the reference point R is defined by

$$[I_R] = -\sum_{i=1}^{N} m_i [R_i - R][R_i - R],$$

where the matrix $[R_i–R]$ is the skew symmetric matrix constructed from the relative position vector $R_i–R$.

Force-torque Equations

Using the centre of mass and inertia matrix, the force and torque equations for a single rigid body take the form

$$F = ma, \quad T = [I_R]\alpha + \omega \times [I_R]\omega,$$

and are known as Newton's second law of motion for a rigid body.

The dynamics of an interconnected system of rigid bodies, B_i, j=1,...,*M*, is formulated by isolating each rigid body and introducing the interaction forces. The resultant of the external and interaction forces on each body, yields the force-torque equations

$$F_j = m_j a_j, \quad T_j = [I_R]_j \alpha_j + \omega_j \times [I_R]_j \omega_j, \quad j = 1, \ldots, M.$$

Newton's formulation yields 6*M* equations that define the dynamics of a system of *M* rigid bodies.

Virtual Work of Forces Acting on a Rigid Body

An alternate formulation of rigid body dynamics that has a number of convenient features is obtained by considering the virtual work of forces acting on a rigid body.

The virtual work of forces acting at various points on a single rigid body can be calculated using the velocities of their point of application and the resultant force and torque. To see this, let the forces F_1, F_2...

F_n act on the points R_1, R_2... R_n in a rigid body. The trajectories of R_i, i=1,...,n are defined by the movement of the rigid body. The velocity of the points R_i along their trajectories are

$$V_i = \vec{\omega} \times (R_i - R) + V,$$

where ω is the angular velocity vector of the body.

Virtual Work

Virtual work is computed from the dot product of each force with the virtual displacement of its point of application

$$\delta W = \sum_{i=1}^{n} F_i \cdot \delta r_i.$$

If the trajectory of a rigid body is defined by a set of generalized coordinatesq$_j$, j=1,..., m, then the virtual displacements dr_i are given by

$$\delta r_i = \sum_{j=1}^{m} \frac{\partial r_i}{\partial q_j} \delta q_j = \sum_{j=1}^{m} \frac{\partial V_i}{\partial \dot{q}_j} \delta q_j.$$

The virtual work of this system of forces acting on the body in terms of the generalized coordinates becomes

$$\delta W = F_1 \cdot (\sum_{j=1}^{m} \frac{\partial V_1}{\partial \dot{q}_j} \delta q_j) + \ldots + F_n \cdot (\sum_{j=1}^{m} \frac{\partial V_n}{\partial \dot{q}_j} \delta q_j)$$

or collecting the coefficients of δq_j

$$\delta W = (\sum_{i-1}^{n} F_i \cdot \frac{\partial V_i}{\partial \dot{q}_1}) \delta q_1 + \ldots + (\sum_{1=1}^{n} F_i \cdot \frac{\partial V_i}{\partial \dot{q}_m}) \delta q_m.$$

Generalized Forces

For simplicity consider a trajectory of a rigid body that is specified by a single generalized coordinate q, such as a rotation angle, then the formula becomes

$$\delta W = (\sum_{i=1}^{n} F_i \cdot \frac{\partial V_i}{\partial \dot{q}}) \delta q = (\sum_{i=1}^{n} F_i \cdot \frac{\partial (\vec{\omega} \times (R_i - R) + V)}{\partial \dot{q}}) \delta q.$$

Introduce the resultant force F and torque T so this equation takes the form

$$\delta W = (F \cdot \frac{\partial V}{\partial \dot{q}} + T \cdot \frac{\partial \vec{\omega}}{\partial \dot{q}}) \delta q.$$

The quantity Q defined by

$$Q = F \cdot \frac{\partial V}{\partial \dot{q}} + T \cdot \frac{\partial \vec{\omega}}{\partial \dot{q}},$$

is known as the generalized force associated with the virtual displacement dq. This formula generalizes to the movement of a rigid body defined by more than one generalized coordinate, that is

$$\delta W = \sum_{j=1}^{m} Q_j \delta q_j,$$

where

$$Q_j = \mathrm{F}\cdot\frac{\partial \mathrm{V}}{\partial \dot{q}_j} + \mathrm{T}\cdot\frac{\partial \vec{\omega}}{\partial \dot{q}_j}, \quad j = 1,\ldots,m.$$

It is useful to note that conservative forces such as gravity and spring forces are derivable from a potential function $V(q_1,\ldots, q_n)$, known as a potential energy. In this case the generalized forces are given by,

$$Q_j = -\frac{\partial V}{\partial q_j}, \quad j = 1,\ldots,m.$$

D'Alembert's form of the Principle of Virtual Work

The equations of motion for a mechanical system of rigid bodies can be determined using D'Alembert's form of the principle of virtual work. The principle of virtual work is used to study the static equilibrium of a system of rigid bodies, however by introducing acceleration terms in Newton's laws this approach is generalized to define dynamic equilibrium.

Static Equilibrium

The static equilibrium of a mechanical system rigid bodies is defined by the condition that the virtual work of the applied forces is zero for any virtual displacement of the system. This is known as the *principle of virtual work.* This is equivalent to the requirement that the generalized forces for any virtual displacement are zero, that is Q_i=0.

Let a mechanical system be constructed from n rigid bodies, B_i, i=1,...,n, and let the resultant of the applied forces on each body be the force-torque pairs, F_i and T_i, i=1,...,n. Notice that these applied forces do not include the reaction forces where the bodies are connected. Finally, assume that the velocity V_i and angular velocities ω_i, i=,1...,n, for each rigid body, are defined by a single generalized coordinate q. Such a system of rigid bodies is said to have one degree of freedom.

The virtual work of the forces and torques, F_i and T_i, applied to this one degree of freedom system is given by

$$\delta W = \sum_{i=1}^{n} (\mathrm{F}_i \cdot\frac{\partial \mathrm{V}_i}{\partial \dot{q}} + \mathrm{T}_i \cdot\frac{\partial \vec{\omega}_i}{\partial \dot{q}})\delta q = Q\delta q,$$

where

$$Q = \sum_{i=1}^{n} (\mathrm{F}_i \cdot \frac{\partial \mathrm{V}_i}{\partial \dot{q}} + \mathrm{T}_i \cdot \frac{\partial \vec{\omega}_i}{\partial \dot{q}}),$$

is the generalized force acting on this one degree of freedom system.

If the mechanical system is defined by m generalized coordinates, q_j, j=1,...,m, then the system has m degrees of freedom and the virtual work is given by,

$$\delta W = \sum_{j=1}^{m} Q_j \delta q_j,$$

where

$$Q_j = \sum_{i=1}^{n} (\mathrm{F}_i \cdot \frac{\partial \mathrm{V}_i}{\partial \dot{q}_j} + \mathrm{T}_i \cdot \frac{\partial \vec{\omega}_i}{\partial \dot{q}_j}), \quad j = 1, \ldots, m.$$

is the generalized force associated with the generalized coordinate q_j. The principle of virtual work states that static equilibrium occurs when these generalized forces acting on the system are zero, that is

$$Q_j = 0, \quad j = 1, \ldots, m.$$

These m equations define the static equilibrium of the system of rigid bodies.

Generalized Inertia Forces

Consider a single rigid body which moves under the action of a resultant for F and torque T, with one degree of freedom defined by the generalized coordinate q. Assume the reference point for the resultant force and torque is the centre of mass of the body, then the generalized inertia force Q* associated with the generalized coordinate q is given by

$$Q^* = -(M\mathrm{A}) \cdot \frac{\partial \mathrm{V}}{\partial \dot{q}} - ([I_R]\alpha + \omega \times [I_R]\omega) \cdot \frac{\partial \vec{\omega}}{\partial \dot{q}}.$$

This inertia force can be computed from the kinetic energy of the rigid body,

$$T = \frac{1}{2} M\mathrm{V} \cdot \mathrm{V} + \frac{1}{2} \vec{\omega} \cdot [I_R] \vec{\omega},$$

by using the formula

$$Q^* = -\left(\frac{d}{dt} \frac{\partial T}{\partial \dot{q}} - \frac{\partial T}{\partial q} \right).$$

A system of n rigid bodies with m generalized coordinates has the kinetic energy

$$T = \sum_{i=1}^{n} \left(\frac{1}{2} M V_i \cdot V_i + \frac{1}{2} \vec{\omega}_i \cdot [I_R] \vec{\omega}_i\right),$$

which can be used to calculate the m generalized inertia forces

$$Q_j^* = -\left(\frac{d}{dt}\frac{\partial T}{\partial \dot{q}_j} - \frac{\partial T}{\partial q_j}\right), \quad j = 1,\ldots,m.$$

Dynamic Equilibrium

D'Alembert's form of the principle of virtual work states that a system of rigid bodies is in dynamic equilibrium when the virtual work of the sum of the applied forces and the inertial forces is zero for any virtual displacement of the system. Thus, dynamic equilibrium of a system of n rigid bodies with m generalized coordinates requires that

$$\delta W = (Q_1 + Q_1^*)\delta q_1 + \ldots + (Q_m + Q_m^*)\delta q_m = 0,$$

for any set of virtual displacements δq_j. This condition yields m equations,

$$Q_j + Q_j^* = 0, \quad j = 1,\ldots,m,$$

which can also be written as

$$\frac{d}{dt}\frac{\partial T}{\partial \dot{q}_j} - \frac{\partial T}{\partial q_j} = Q_j, \quad j = 1,\ldots,m.$$

The result is a set of m equations of motion that define the dynamics of the rigid body system.

Lagrange's Equations

If the generalized forces Q_j are derivable from a potential energy $V(q_1,\ldots,q_m)$, then these equations of motion take the form

$$\frac{d}{dt}\frac{\partial T}{\partial \dot{q}_j} - \frac{\partial T}{\partial q_j} = -\frac{\partial V}{\partial q_j}, \quad j = 1,\ldots,m.$$

In this case, introduce the Lagrangian, L=T-V, so these equations of motion become

$$\frac{d}{dt}\frac{\partial L}{\partial \dot{q}_j} - \frac{\partial L}{\partial q_j} = 0 \quad j = 1,\ldots,m.$$

These are known as Lagrange's equations of motion.

Linear and Angular Momentum

System of Particles: The linear and angular momentum of a rigid system of particles is formulated by measuring the position and velocity of the particles relative to the centre of mass. Let the system of particles P_i, i=1,...,n be located at the coordinates r_i and velocities v_i. Select a reference point R and compute the relative position and velocity vectors,

$$r_i = (r_i - R) + R, \quad v_i = \frac{d}{dt}(r_i - R) + V.$$

The total linear and angular momentum vectors relative to the reference point R are

$$p = \frac{d}{dt}\left(\sum_{i=1}^{n} m_i (r_i - R)\right) + \left(\sum_{i=1}^{n} m_i\right)V,$$

and

$$L = \sum_{i=1}^{n} m_i (r_i - R) \times \frac{d}{dt}(r_i - R) + \left(\sum_{i=1}^{n} m_i (r_i - R)\right) \times V.$$

If R is chosen as the centre of mass these equations simplify to

$$p = MV, \quad L = \sum_{i=1}^{n} m_i (r_i - R) \times \frac{d}{dt}(r_i - R).$$

Rigid System of Particles

To specialize these formulas to a rigid body, assume the particles are rigidly connected to each other so P_i, i=1,...,n are located by the coordinates r_i and velocities v_i. Select a reference point R and compute the relative position and velocity vectors,

$$r_i = (r_i - R) + R, \quad v_i = \omega \times (r_i - R) + V,$$

where ω is the angular velocity of the system.

The linear momentum and angular momentum of this rigid system measured relative to the centre of mass R is

$$p = \left(\sum_{i=1}^{n} m_i\right)V, \quad L = \sum_{i=1}^{n} m_i (r_i - R) \times v_i = \sum_{i=1}^{n} m_i (r_i - R) \times (\omega \times (r_i - R)).$$

These equations simplify to become,

$$p = MV, \quad L = [I_R]\omega,$$

where M is the total mass of the system and $[I_R]$ is the moment of inertia matrix defined by

$$[I_R] = -\sum_{i=1}^{n} m_i [r_i - R][r_i - R],$$

where $[r_i\text{-}R]$ is the skew-symmetric matrix constructed from the vector $r_i\text{-}R$.

Applications

- For the analysis of robotic systems
- For the biomechanical analysis of animals, humans or humanoid systems
- For the analysis of space objects
- for the design and development of dynamics-based sensors like gyroscopic sensors etc.
- For the design and development of various stability enhancement applications in automobiles etc.
- For improving the graphics of video games which involves rigid bodies.

4

Mechanical Vibrations

Introduction

Vibration is a mechanical phenomenon whereby oscillations occur about an equilibrium point. The oscillations may be periodic such as the motion of a pendulum or random such as the movement of a tire on a gravel road. Vibration is occasionally "desirable". For example the motion of a tuning fork, the reed in a woodwind instrument or harmonica, or mobile phones or the cone of a loudspeaker is desirable vibration, necessary for the correct functioning of the various devices.

More often, vibration is undesirable, wasting energy and creating unwanted sound – noise. For example, the vibrational motions of engines, electric motors, or any mechanical device in operation are typically unwanted. Such vibrations can be caused by imbalances in the rotating parts, uneven friction, the meshing of gear teeth, etc. Careful designs usually minimize unwanted vibrations. The study of sound and vibration are closely related. Sound, or "pressure waves", are generated by vibrating structures (e.g. vocal cords); these pressure waves can also induce the vibration of structures (e.g. ear drum). Hence, when trying to reduce noise it is often a problem in trying to reduce vibration.

Types of Vibration

Free vibration occurs when a mechanical system is set off with an initial input and then allowed to vibrate freely. Examples of this type of vibration are pulling a child back on a swing and then letting go or hitting a tuning fork and letting it ring. The mechanical system will then vibrate at one or more of its "natural frequency" and damp down to zero.

Forced vibration is when a time-varying disturbance (load, displacement or velocity) is applied to a mechanical system. The disturbance can be a periodic, steady-state input, a transient input, or a random input. The periodic input can be a harmonic or a non-harmonic disturbance. Examples of these types of vibration include a shaking washing machine due to an imbalance, transportation vibration (caused by truck engine, springs, road, etc.), or the vibration of a building during an earthquake. For linear systems, the frequency of the steady-state vibration response resulting from the application of a periodic, harmonic input is equal to the frequency of the applied force or motion, with the response magnitude being dependent on the actual mechanical system.

Vibration Testing

Vibration testing is accomplished by introducing a forcing function into a structure, usually with some type of shaker. Alternately, a DUT (device under test) is attached to the "table" of a shaker. Vibration testing is performed to examine the response of a device under test (DUT) to a defined vibration environment. The measured response may be fatigue life, resonant frequencies or squeak and rattle sound output (NVH). Squeak and rattle testing is performed with a special type of quiet shaker that produces very low sound levels while under operation.

For relatively low frequency forcing, servohydraulic (electrohydraulic) shakers are used. For higher frequencies, electrodynamic shakers are used. Generally, one or more "input" or "control" points located on the DUT-side of a fixture is kept at a specified acceleration. Other "response" points experience maximum vibration level (resonance) or minimum vibration level (anti-resonance). It is often desirable to achieve anti-resonance in order to keep a system from becoming too noisy, or to reduce strain on certain parts of a system due to vibration modes caused by specific frecquencies of vibration.

Two typical types of vibration tests performed are random- and sine test. Sine (one-frequency-at-a-time) tests are performed to survey the structural response of the device under test (DUT). A random (all frequencies at once) test is generally considered to more closely replicate a real world environment, such as road inputs to a moving automobile.

Most vibration testing is conducted in a single DUT axis at a time, even though most real-world vibration occurs in various axes simultaneously. MIL-STD-810G, released in late 2008, Test Method 527, calls for multiple exciter testing. The vibration test fixture which is used to attach the DUT to the shaker table must be designed for the frequency range of the vibration test spectrum. Generally for smaller fixtures and lower frequency ranges, the designer targets a fixture design which is free of resonances in the test frequency range. This becomes more difficult as the DUT gets larger and as the test frequency increases. In these cases multi-point control strategies can be employed to mitigate some of the resonances which may be present in the fixture. Devices specifically designed to trace or record vibrations are called vibroscopes.

Vibration Analysis

The fundamentals of vibration analysis can be understood by studying the simple mass–spring–damper model. Indeed, even a complex structure such as an automobile body can be modelled as a "summation" of simple mass–spring–damper models. The mass–spring–damper model is an example of a simple harmonic oscillator. The mathematics used to describe its behaviour is identical to other simple harmonic oscillators such as the RLC circuit.

Note: In this article the step by step mathematical derivations will not be included, but will focus on the major equations and concepts in vibration analysis. Please refer to the references at the end of the article for detailed derivations.

Free Vibration without Damping

To start the investigation of the mass–spring–damper we will assume the damping is negligible and that there is no external force applied to the mass (i.e. free vibration). The force applied to the mass by the spring is proportional to the amount the spring is stretched "x" (we will assume the spring is already compressed due to the weight of the mass). The proportionality constant, k, is the stiffness of the spring and has units of force/distance (e.g. lbf/in or N/m). The negative sign indicates that the force is always opposing the motion of the mass attached to it:

$$F_s = -kx.$$

The force generated by the mass is proportional to the acceleration of the mass as given by Newton's second law of motion :

$$\Sigma F = ma = m\ddot{x} = m\frac{d^2x}{dt^2}.$$

The sum of the forces on the mass then generates this ordinary differential equation: $m\ddot{x} + kx = 0.$

If we assume that we start the system to vibrate by stretching the spring by the distance of A and letting go, the solution to the above equation that describes the motion of mass is:

$$x(t) = A\cos(2\pi f_n t).$$

This solution says that it will oscillate with simple harmonic motion that has an amplitude of A and a frequency of f_n. The number f_n is one of the most important quantities in vibration analysis and is called the undamped natural frequency. For the simple mass–spring system, f_n is defined as:

$$f_n = \frac{1}{2\pi}\sqrt{\frac{k}{m}}.$$

Note: Angular frequency ω ($\omega = 2\pi f$) with the units of radians per second is often used in equations because it simplifies the equations, but is normally converted to "standard" frequency (units of Hz or equivalently cycles per second) when stating the frequency of a system. If you know the mass and stiffness of the system you can determine the frequency at which the system will vibrate once it is set in motion by an initial disturbance using the above stated formula. Every vibrating system has one or more natural frequencies that it will vibrate at once it is disturbed. This simple relation can be used to understand in general what will happen to a more complex system once we add mass or stiffness. For example, the above formula explains why when a car or truck is fully loaded the suspension will feel 3 softer3 than unloaded because the mass has increased and therefore reduced the natural frequency of the system.

What Causes the System to Vibrate: from Conservation of Energy Point of View

Vibrational motion could be understood in terms of conservation of energy. In the above example we have extended the spring by a value of x and therefore have stored some potential energy ($\frac{1}{2}kx^2$) in the spring. Once we let go of the spring, the spring tries to return to its un-stretched state (which is the minimum potential energy state) and in the process accelerates the mass. At the point where the spring has reached its un-stretched state all the potential energy that we supplied by stretching it has been transformed into kinetic energy ($\frac{1}{2}mv^2$). The mass then begins to decelerate because it is now compressing the spring and in the process transferring the kinetic energy back to its potential. Thus oscillation of the spring amounts to the transferring back and forth of the kinetic energy into potential energy. In our simple model the mass will continue to oscillate forever at the same magnitude, but in a real system there is always something called damping that dissipates the energy, eventually bringing it to rest.

Free Vibration with Damping

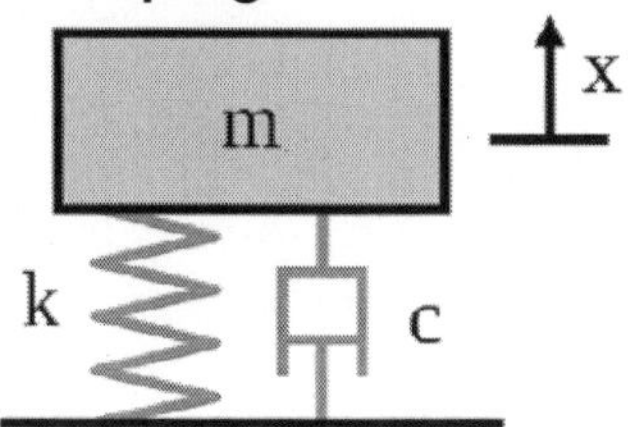

Figure: *Mass Spring Damper Model*

We now add a "viscous" damper to the model that outputs a force that is proportional to the velocity of the mass. The damping is called viscous because it models the effects of an object within a fluid. The proportionality constant *c* is called the damping coefficient and has units of Force over velocity (lbf s/ in or *N* s/m).

$$F_d = -cv = -c\dot{x} = -c\frac{dx}{dt}.$$

By summing the forces on the mass we get the following ordinary differential equation:

$$m\ddot{x} + c\dot{x} + kx = 0.$$

The solution to this equation depends on the amount of damping. If the damping is small enough the system will still vibrate, but eventually, over time, will stop vibrating. This case is called underdamping – this case is of most interest in vibration analysis. If we increase the damping just to the point where the system no longer oscillates we reach the point of critical damping (if the damping is increased past critical damping the system is called overdamped). The value that the damping coefficient needs to reach for critical damping in the mass spring damper model is:

$$c_c = 2\sqrt{km}.$$

To characterize the amount of damping in a system a ratio called the damping ratio (also known as damping factor and % critical damping) is used. This damping ratio is just a ratio of the actual damping over the amount of damping required to reach critical damping. The formula for the damping ratio (ζ) of the mass spring damper model is:

$$\zeta = \frac{c}{2\sqrt{km}}.$$

For example, metal structures (e.g. airplane fuselage, engine crankshaft) will have damping factors less than 0.05 while automotive suspensions in the range of 0.2–0.3.

The solution to the underdamped system for the mass spring damper model is the following:

$$x(t) = Xe^{-\zeta\omega_n t}\cos(\sqrt{1-\zeta^2}\,\omega_n t - \phi), \qquad \omega_n = 2\pi f_n.$$

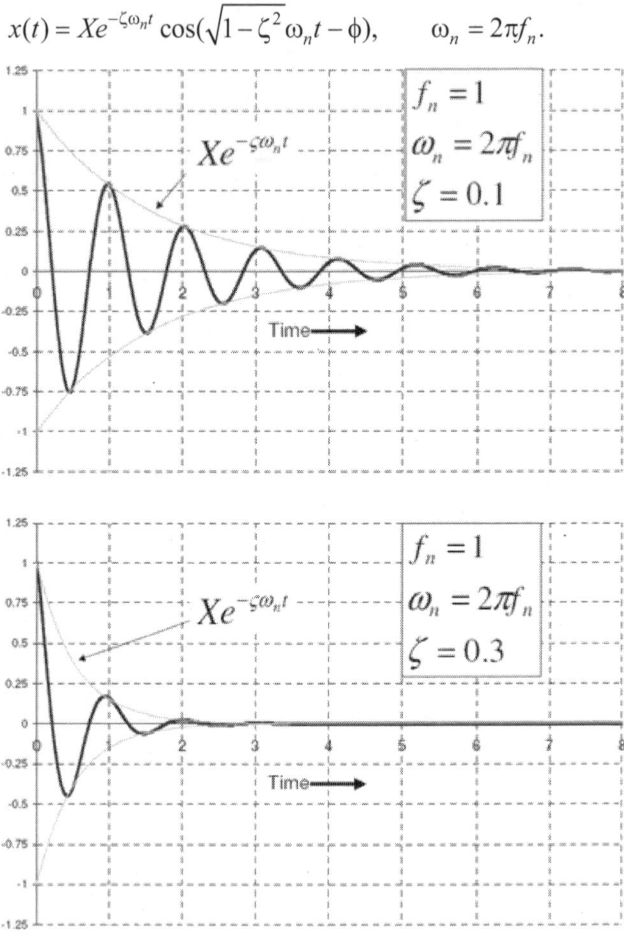

The value of X, the initial magnitude, and ϕ, the phase shift, are determined by the amount the spring is stretched. The formulas for these values can be found in the references.

Damped and Undamped Natural Frequencies

The major points to note from the solution are the exponential term and the cosine function. The exponential term defines how quickly the system "damps" down – the larger the damping ratio, the quicker it damps to zero. The cosine function is the oscillating portion of the solution, but the frequency of the oscillations is different from the undamped case.

The frequency in this case is called the "damped natural frequency", f_d, and is related to the undamped natural frequency by the following formula:

$$f_d = f_n\sqrt{1-\zeta^2}.$$

The damped natural frequency is less than the undamped natural frequency, but for many practical cases the damping ratio is relatively small and hence the difference is negligible. Therefore the damped and undamped description are often dropped when stating the natural frequency (e.g. with 0.1 damping ratio, the damped natural frequency is only 1% less than the undamped).

The plots to the side present how 0.1 and 0.3 damping ratios effect how the system will "ring" down over time. What is often done in practice is to experimentally measure the free vibration after an impact (for example by a hammer) and then determine the natural frequency of the system by measuring the rate of oscillation as well as the damping ratio by measuring the rate of decay. The natural frequency and damping ratio are not only important in free vibration, but also characterize how a system will behave under forced vibration.

Forced Vibration with Damping

The behaviour of the spring mass damper model when we add a harmonic force. A force of this type could, for example, be generated by a rotating imbalance.

$$F = F_0 \sin(2\pi ft).$$

If we again sum the forces on the mass we get the following ordinary differential equation:

$$m\ddot{x} + c\dot{x} + kx = F_0 \sin(2\pi ft).$$

The steady state solution of this problem can be written as:

$$x(t) = X \sin(2\pi ft + \phi).$$

The result states that the mass will oscillate at the same frequency, f, of the applied force, but with a phase shift ϕ.

The amplitude of the vibration "X" is defined by the following formula.

$$X = \frac{F_0}{k}\frac{1}{\sqrt{(1-r^2)^2 + (2\zeta r)^2}}.$$

Where "r" is defined as the ratio of the harmonic force frequency over the undamped natural frequency of the mass–spring–damper model.

$$r = \frac{f}{f_n}.$$

The phase shift, ϕ, is defined by the following formula.

$$\phi = \arctan\left(\frac{2\zeta r}{1-r^2}\right).$$

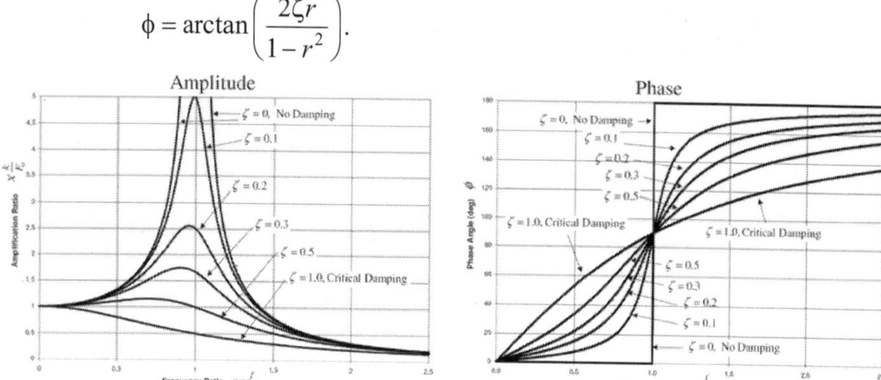

The plot of these functions, called "the frequency response of the system", presents one of the most important features in forced vibration. In a lightly damped system when the forcing frequency nears the natural frequency ($r \approx 1$) the amplitude of the vibration can get extremely high. This phenomenon is called resonance (subsequently the natural frequency of a system is often referred to as the resonant frequency). In rotor bearing systems any rotational speed that excites a resonant frequency is referred to as a critical speed.

If resonance occurs in a mechanical system it can be very harmful – leading to eventual failure of the system. Consequently, one of the major reasons for vibration analysis is to predict when this type of resonance may occur and then to determine what steps to take to prevent it from occurring. As the amplitude plot shows, adding damping can significantly reduce the magnitude of the vibration. Also, the magnitude can be reduced if the natural frequency can be shifted away from the forcing frequency by changing the stiffness or mass of the system. If the system cannot be changed, perhaps the forcing frequency can be shifted (for example, changing the speed of the machine generating the force).

The following are some other points in regards to the forced vibration shown in the frequency response plots.

- At a given frequency ratio, the amplitude of the vibration, X, is directly proportional to the amplitude of the force F_0 (e.g. if you double the force, the vibration doubles)

- With little or no damping, the vibration is in phase with the forcing frequency when the frequency ratio $r < 1$ and 180 degrees out of phase when the frequency ratio $r > 1$
- When $r \ll 1$ the amplitude is just the deflection of the spring under the static force F_0. This deflection is called the static deflection δ_{st}. Hence, when $r \ll 1$ the effects of the damper and the mass are minimal.
- When $r \gg 1$ the amplitude of the vibration is actually less than the static deflection δ_{st}. In this region the force generated by the mass ($F = ma$) is dominating because the acceleration seen by the mass increases with the frequency. Since the deflection seen in the spring, X, is reduced in this region, the force transmitted by the spring ($F = kx$) to the base is reduced. Therefore the mass–spring–damper system is isolating the harmonic force from the mounting base – referred to as vibration isolation. Interestingly, more damping actually reduces the effects of vibration isolation when $r \gg 1$ because the damping force ($F = cv$) is also transmitted to the base.
- whatever the damping is, the vibration is 90 degrees out of phase with the forcing frequency when the frequency ratio r =1,which is very helpful when it comes to determining the natural frequency of the system.
- whatever the damping is, when r $\gg$ 1, the vibration is 180 degrees out of phase with the forcing frequency
- whatever the damping is, when r $\ll$ 1, the vibration is in phase with the forcing frequency

What Causes Resonance?

Resonance is simple to understand if you view the spring and mass as energy storage elements – with the mass storing kinetic energy and the spring storing potential energy. As discussed earlier, when the mass and spring have no external force acting on them they transfer energy back and forth at a rate equal to the natural frequency. In other words, if energy is to be efficiently pumped into both the mass and spring the energy source needs to feed the energy in at a rate equal to the natural frequency. Applying a force to the mass and spring is similar to pushing a child on swing, you need to push at the correct moment if you want the swing to get higher and higher. As in the case of the swing, the force applied does not necessarily have to be high to get large motions; the pushes just need to keep adding energy into the system.

The damper, instead of storing energy, dissipates energy. Since the damping force is proportional to the velocity, the more the motion, the more the damper dissipates the energy. Therefore a point will come when the energy dissipated by the damper will equal the energy being fed in by the force. At this point, the system has reached its maximum amplitude and will continue to vibrate at this level as long as the force applied stays the same. If no damping exists, there is nothing to dissipate the energy and therefore theoretically the motion will continue to grow on into infinity.

Applying "complex" Forces to the Mass–spring–damper Model

In a previous section only a simple harmonic force was applied to the model, but this can be extended considerably using two powerful mathematical tools. The first is the Fourier transform that takes a signal as a function of time (time domain) and breaks it down into its harmonic components as a function of frequency (frequency domain). For example, let us apply a force to the mass–spring–damper model that repeats the following cycle – a force equal to 1 newton for 0.5 second and then no force for 0.5 second. This type of force has the shape of a 1 Hz square wave.

The Fourier transform of the square wave generates a frequency spectrum that presents the magnitude of the harmonics that make up the square wave (the phase is also generated, but is typically of less concern and therefore is often not plotted). The Fourier transform can also be used to analyze non-periodic functions such as transients (e.g. impulses) and random functions. With the advent of the modern computer the Fourier transform is almost always computed using the Fast Fourier Transform (FFT) computer algorithm in combination with a window function.

In the case of our square wave force, the first component is actually a constant force of 0.5 newton and is represented by a value at "0" Hz in the frequency spectrum. The next component is a 1 Hz sine wave with an amplitude of 0.64. This is shown by the line at 1 Hz. The remaining components are at odd frequencies and it takes an infinite amount of sine waves to generate the perfect square wave. Hence, the Fourier transform allows you to interpret the force as a sum of sinusoidal forces being applied instead of a more "complex" force (e.g. a square wave).

In the previous section, the vibration solution was given for a single harmonic force, but the Fourier transform will in general give multiple harmonic forces. The second mathematical tool, "the principle of superposition", allows you to sum the solutions from multiple forces if the system is linear. In the case of the spring–mass–damper model, the system is linear if the spring force is proportional to the displacement and the damping is proportional to the velocity over the range of motion of interest. Hence, the solution to the problem with a square wave is summing the predicted vibration from each one of the harmonic forces found in the frequency spectrum of the square wave.

Frequency Response Model

We can view the solution of a vibration problem as an input/output relation – where the force is the input and the output is the vibration. If we represent the force and vibration in the frequency domain (magnitude and phase) we can write the following relation:

$$X(i\omega) = H(i\omega)\cdot F(i\omega) \quad or \quad H(i\omega) = \frac{X(i\omega)}{F(i\omega)}.$$

$H(i\omega)$ is called the frequency response function (also referred to as the transfer function, but not technically as accurate) and has both a magnitude and phase component (if represented as a complex number, a real and imaginary component). The magnitude of the frequency response function (FRF) was presented earlier for the mass–spring–damper system.

$$|H(i\omega)| = \left|\frac{X(i\omega)}{F(i\omega)}\right| = \frac{1}{k}\frac{1}{\sqrt{(1-r^2)^2+(2\zeta r)^2}}, \quad \text{where} \quad r = \frac{f}{f_n} = \frac{\omega}{\omega_n}.$$

The phase of the FRF was also presented earlier as:

$$\angle H(i\omega) = -\arctan\left(\frac{2\zeta r}{1-r^2}\right).$$

For example, let us calculate the FRF for a mass–spring–damper system with a mass of 1 kg, spring stiffness of 1.93 N/mm and a damping ratio of 0.1. The values of the spring and mass give a natural frequency of 7 Hz for this specific system. If we apply the 1 Hz square wave from earlier we can calculate the predicted vibration of the mass. The figure illustrates the resulting vibration. It happens in this example that the fourth harmonic of the square wave falls at 7 Hz. The frequency response of the mass–spring–damper therefore outputs a high 7 Hz vibration even though the input force had a relatively low 7 Hz harmonic. This example highlights that the resulting vibration is dependent on both the forcing function and the system that the force is applied to. This is done by performing an inverse Fourier Transform that converts frequency domain data to time domain. In practice, this is rarely done because the frequency spectrum provides all the necessary information.

The frequency response function (FRF) does not necessarily have to be calculated from the knowledge of the mass, damping, and stiffness of the system, but can be measured experimentally. For example, if you apply a known force and sweep the frequency and then measure the resulting vibration you can calculate the frequency response function and then characterize the system. This technique is used in the field of experimental modal analysis to determine the vibration characteristics of a structure.

Multiple Degrees of Freedom Systems and Mode Shapes

The simple mass–spring damper model is the foundation of vibration analysis, but what about more complex systems? The mass–spring–damper model described above is called a single degree of freedom (SDOF) model since we have assumed the mass only moves up and down. In the case of more complex systems we need to discretize the system into more masses and allow them to move in more than one direction – adding degrees of freedom. The major concepts of multiple degrees of freedom (MDOF) can be understood by looking at just a 2 degree of freedom model as shown in the figure.

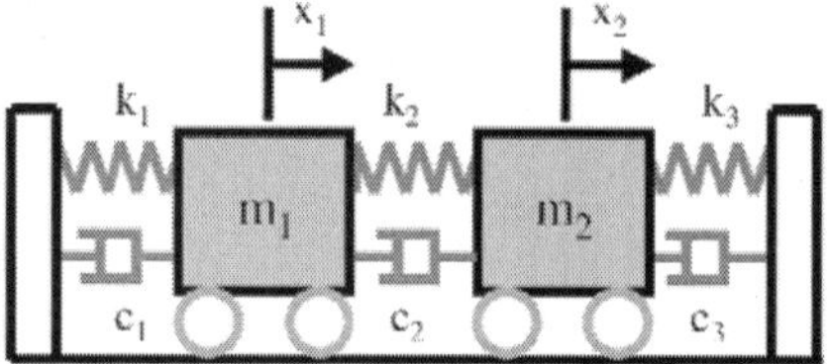

Figure: *Degree of freedom model*

The equations of motion of the 2DOF system are found to be:

$$m_1 \ddot{x}_1 + (c_1 + c_2)\dot{x}_1 - c_2 \dot{x}_2 + (k_1 + k_2)x_1 - k_2 x_2 = f_1,$$

$$m_2 \ddot{x}_2 - c_2 \dot{x}_1 + (c_2 + c_3)\dot{x}_2 - k_2 x_1 + (k_2 + k_3)x_2 = f_2.$$

We can rewrite this in matrix format:

$$\begin{bmatrix} m_1 & 0 \\ 0 & m_2 \end{bmatrix} \begin{Bmatrix} \ddot{x}_1 \\ \ddot{x}_2 \end{Bmatrix} + \begin{bmatrix} c_1 + c_2 & -c_2 \\ -c_2 & c_2 + c_3 \end{bmatrix} \begin{Bmatrix} \dot{x}_1 \\ \dot{x}_2 \end{Bmatrix} + \begin{bmatrix} k_1 + k_2 & -k_2 \\ -k_2 & k_2 + k_3 \end{bmatrix} \begin{Bmatrix} x_1 \\ x_2 \end{Bmatrix} = \begin{Bmatrix} f_1 \\ f_2 \end{Bmatrix}.$$

A more compact form of this matrix equation can be written as:

$$[M]\{\ddot{x}\} + [C]\{\dot{x}\} + [K]\{x\} - \{f\}$$

where $[M]$, $[C]$, and $[K]$ are symmetric matrices referred respectively as the mass, damping, and stiffness matrices. The matrices are NxN square matrices where N is the number of degrees of freedom of the system.

In the following analysis we will consider the case where there is no damping and no applied forces (i.e. free vibration). The solution of a viscously damped system is somewhat more complicated.

$$[M]\{\ddot{x}\}+[K]\{x\}=0.$$

This differential equation can be solved by assuming the following type of solution:

$$\{x\}=\{X\}e^{i\omega t}.$$

Note: Using the exponential solution of $\{X\}e^{i\omega t}$ is a mathematical trick used to solve linear differential equations. If we use Euler's formula and take only the real part of the solution it is the same cosine solution for the 1 DOF system. The exponential solution is only used because it easier to manipulate mathematically.

The equation then becomes:

$$\left[-\omega^2[M]+[K]\right]\{X\}e^{i\omega t}=0.$$

Since $e^{i\omega t}$ cannot equal zero the equation reduces to the following.

$$\left[[K]-\omega^2[M]\right]\{X\}=0.$$

Eigenvalue Problem

This is referred to an eigenvalue problem in mathematics and can be put in the standard format by pre-multiplying the equation by $[M]^{-1}$

$$\left[[M]^{-1}[K]-\omega^2[M]^{-1}[M]\right]\{X\}=0$$

and if we let $[M]^{-1}[K]=[A]$ and $\lambda=\omega^2$

$$\left[[A]-\lambda[I]\right]\{X\}=0.$$

The solution to the problem results in N eigenvalues (i.e. $\omega_1^2,\omega_2^2,\cdots\omega_N^2$), where N corresponds to the number of degrees of freedom. The eigenvalues provide the natural frequencies of the system. When these eigenvalues are substituted back into the original set of equations, the values of $\{X\}$ that correspond to each eigenvalue are called the eigenvectors. These eigenvectors represent the mode shapes of the system. The solution of an eigenvalue problem can be quite cumbersome (especially for problems with many degrees of freedom), but fortunately most math analysis programmes have eigenvalue routines.

The eigenvalues and eigenvectors are often written in the following matrix format and describe the modal model of the system:

$$\left[\diagdown \omega_r^2 \diagdown\right] = \begin{bmatrix} \omega_1^2 & \cdots & 0 \\ \vdots & \ddots & \vdots \\ 0 & \cdots & \omega_N^2 \end{bmatrix} \text{and} \left[\Psi\right] = \left[\{\psi_1\}\{\psi_2\}\cdots\{\psi_N\}\right].$$

A simple example using our 2 DOF model can help illustrate the concepts. Let both masses have a mass of 1 kg and the stiffness of all three springs equal 1000 N/m. The mass and stiffness matrix for this problem are then:

$$[M] = \begin{bmatrix} 1 & 0 \\ 0 & 1 \end{bmatrix} \text{and} [K] = \begin{bmatrix} 2000 & -1000 \\ -1000 & 2000 \end{bmatrix}.$$

Then $[A] = \begin{bmatrix} 2000 & -1000 \\ -1000 & 2000 \end{bmatrix}.$

The eigenvalues for this problem given by an eigenvalue routine will be:

$$\left[\diagdown \omega_r^2 \diagdown\right] = \begin{bmatrix} 1000 & 0 \\ 0 & 3000 \end{bmatrix}.$$

The natural frequencies in the units of hertz are then (remembering $\omega = 2\pi f$) $f_1 = 5.033$ Hz and $f_2 = 8.717$ Hz .

The two mode shapes for the respective natural frequencies are given as:

$$[\Psi] = \left[\{\psi_1\}\{\psi_2\}\right] = \left[\begin{Bmatrix} -0.707 \\ -0.707 \end{Bmatrix}_1 \begin{Bmatrix} 0.707 \\ -0.707 \end{Bmatrix}_2\right].$$

Since the system is a 2 DOF system, there are two modes with their respective natural frequencies and shapes. The mode shape vectors are not the absolute motion, but just describe relative motion of the degrees of freedom. In our case the first mode shape vector is saying that the masses are moving together in phase since they have the same value and sign. In the case of the second mode shape vector, each mass is moving in opposite direction at the same rate.

Illustration of a Multiple DOF Problem

When there are many degrees of freedom, the best method of visualizing the mode shapes is by animating them. An example of animated mode shapes is shown in the figure below for a cantileveredI-beam. In this case, the finite element method was used to generate an approximation to the mass and stiffness matrices and solve a discrete eigenvalue problem. Note that, in this case, the finite element method provides an approximation of the 3D electrodynamics model (for which there exists an infinite number of vibration modes and frequencies). Therefore, this relatively simple model that has over 100 degrees of freedom and hence as many natural frequencies and mode shapes, provides a good approximation for the first natural frequencies and modes†. Generally, only the first few modes are important for practical applications.

Multiple DOF problem converted to a single DOF problem

The eigenvectors have very important properties called orthogonality properties. These properties can be used to greatly simplify the solution of multi-degree of freedom models. It can be shown that the eigenvectors have the following properties:

$$[\Psi]^T[M][\Psi] = \left[\diagdown m_r \diagdown\right],$$

$$[\Psi]^T[K][\Psi] = \left[\diagdown k_r \diagdown\right].$$

$\left[\diagdown m_r \diagdown\right]$ and $\left[\diagdown k_r \diagdown\right]$ are diagonal matrices that contain the modal mass and stiffness values for each one of the modes. (Note: Since the eigenvectors (mode shapes) can be arbitrarily scaled, the orthogonality properties are often used to scale the eigenvectors so the modal mass value for each mode is equal to 1. The modal mass matrix is therefore an identity matrix)

These properties can be used to greatly simplify the solution of multi-degree of freedom models by making the following coordinate transformation.

$$\{x\} = [\Psi]\{q\}.$$

If we use this coordinate transformation in our original free vibration differential equation we get the following equation.

$$[M][\Psi]\{\ddot{q}\} + [K][\Psi]\{q\} = 0.$$

We can take advantage of the orthogonality properties by premultiplying this equation by $[\Psi]^T$

$$[\Psi]^T[M][\Psi]\{\ddot{q}\} + [\Psi]^T[K][\Psi]\{q\} = 0.$$

The orthogonality properties then simplify this equation to:

$$\left[\diagdown m_r \diagdown\right]\{\ddot{q}\}+\left[\diagdown k_r \diagdown\right]\{q\}=0.$$

This equation is the foundation of vibration analysis for multiple degree of freedom systems. A similar type of result can be derived for damped systems. The key is that the modal and stiffness matrices are diagonal matrices and therefore we have "decoupled" the equations. In other words, we have transformed our problem from a large unwieldy multiple degree of freedom problem into many single degree of freedom problems that can be solved using the same methods outlined above.

Instead of solving for x we are instead solving for q, referred to as the modal coordinates or modal participation factors.

It may be clearer to understand if we write $\{x\}=[\Psi]\{q\}$ as:

$$\{x_n\}=q_1\{\psi\}_1+q_2\{\psi\}_2+q_3\{\psi\}_3+\cdots+q_N\{\psi\}_N.$$

Written in this form we can see that the vibration at each of the degrees of freedom is just a linear sum of the mode shapes. Furthermore, how much each mode "participates" in the final vibration is defined by q, its modal participation factor.

Resonance and its Effects

Mechanical resonance is the tendency of a mechanical system to respond at greater amplitude when the frequency of its oscillations matches the system's natural frequency of vibration (its *resonance frequency* or *resonant frequency*) than it does at other frequencies. It may cause violent swaying motions and even catastrophic failure in improperly constructed structures including bridges, buildings and airplanes—a phenomenon known as resonance disaster.

Avoiding resonance disasters is a major concern in every building, tower and bridge construction project. The Taipei 101 building relies on a 660-ton pendulum — a tuned mass damper — to modify the response at resonance. Furthermore, the structure is designed to resonate at a frequency which does not typically occur. Buildings in seismic zones are often constructed to take into account the oscillating frequencies of expected ground motion. In addition, engineers designing objects having engines must ensure that the mechanical resonant frequencies of the component parts do not match driving vibrational frequencies of the motors or other strongly oscillating parts.

Many resonant objects have more than one resonance frequency. It will vibrate easily at those frequencies, and less so at other frequencies.

Many clocks keep time by mechanical resonance in a balance wheel, pendulum, or quartz crystal.

Description

The natural frequency of a simple mechanical system consisting of a weight suspended by a spring is:

$$f = \frac{1}{2\pi}\sqrt{\frac{k}{m}}$$

where m is the mass and k is the spring constant.

A swing set is a simple example of a resonant system with which most people have practical experience. It is a form of pendulum. If the system is excited (pushed) with a period between pushes equal to the inverse of the pendulum's natural frequency, the swing will swing higher and higher, but if excited at a different frequency, it will be difficult to move. The resonance frequency of a pendulum, the only frequency at which it will vibrate, is given approximately, for small displacements, by the equation:

$$f = \frac{1}{2\pi}\sqrt{\frac{g}{L}}$$

where g is the acceleration due to gravity (about 9.8 m/s^2 near the surface of Earth), and L is the length from the pivot point to the centre of mass. (An elliptic integral yields a description for any displacement). Note that, in this approximation, the frequency does not depend on mass.

Mechanical resonators work by transferring energy repeatedly from kinetic to potential form and back again. In the pendulum, for example, all the energy is stored as gravitational energy (a form of potential energy) when the bob is instantaneously motionless at the top of its swing. This energy is proportional to both the mass of the bob and its height above the lowest point. As the bob descends and picks up speed, its potential energy is gradually converted to kinetic energy (energy of movement), which is proportional to the bob's mass and to the square of its speed. When the bob is at the bottom of its travel, it has maximum kinetic energy and minimum potential energy. The same process then happens in reverse as the bob climbs towards the top of its swing.

Some resonant objects have more than one resonance frequency, particularly at harmonics (multiples) of the strongest resonance. It will vibrate easily at those frequencies, and less so at other frequencies. It will "pick out" its resonance frequency from a complex excitation, such as an impulse or a wideband noise excitation. In effect, it is filtering out all frequencies other than its resonance. In the example above, the swing cannot easily be excited by harmonic frequencies, but can be excited by subharmonics.

Examples

Various examples of mechanical resonance include:

- musical instruments (acoustic resonance).
- Most clocks keep time by mechanical resonance in a balance wheel, pendulum, or quartz crystal.
- tidal resonance of the Bay of Fundy.
- Orbital resonance as in some moons of the solar system's gas giants.
- The resonance of the basilar membrane in the ear.
- Making a child's swingswing higher by pushing it at each swing.
- A wineglass breaking when someone sings a loud note at exactly the right pitch.

Resonance may cause violent swaying motions in improperly constructed structures, such as bridges and buildings. The London Millennium Footbridge (nicknamed the *Wobbly Bridge*) exhibited this problem. A faulty bridge can even be destroyed by its resonance; that is why soldiers are trained not to march in lockstep across a bridge, although it is suspected to be a myth, MythBusters' 'Breakstep Bridge'. Mechanical systems store potential energy in different forms. For example, a spring/mass system stores energy as tension in the spring, which is ultimately stored as the energy of bonds between atoms.

Resonance Disaster

In mechanics and construction a resonance disaster describes the destruction of a building or a technical mechanism by induced vibrations at a system's resonance frequency, which causes it to oscillate. Periodic excitation optimally transfers to the system the energy of the vibration and stores it there. Because of this repeated storage and additional energy input the system swings ever more strongly, until its load limit is exceeded.

Failure of the Original Tacoma Narrows Bridge

The dramatic, rhythmic twisting that resulted in the 1940 collapse of "Galloping Gertie", the original Tacoma Narrows Bridge, is sometimes characterized in physics textbooks as a classic example of resonance; however, this description is misleading. The catastrophic vibrations that destroyed the bridge were not due to simple mechanical resonance, but to a more complicated oscillation caused by interactions between the bridge and the winds passing through its structure — a phenomenon known as aeroelastic flutter. Robert H. Scanlan, father of the field of bridge aerodynamics, wrote an article about this misunderstanding.

Other Examples

- Collapse of Broughton Suspension Bridge (due to soldiers walking in step)
- Collapse of Angers Bridge
- Collapse of KönigsWusterhausen Central Tower
- Resonance of the Millennium Bridge
- Evacuation of the 39-story TechnoMart commercial-residential high-rise in Korea in 2011 due to a class performing Tae Bo exercises to the song "The Power".

Applications

Various method of inducing mechanical resonance in a medium exist. Mechanical waves can be generated in a medium by subjecting an electromechanical element to an alternating electric field having a frequency which induces mechanical resonance and is below any electrical resonance frequency. Such devices can apply mechanical energy from an external source to an element to mechanically stress the element or apply mechanical energy produced by the element to an external load.

The United States Patent Office classifies devices that tests mechanical resonance under subclass 579, resonance, frequency, or amplitude study, of Class 73, Measuring and testing. This subclass is itself indented under subclass 570, Vibration. Such devices test an article or mechanism by subjecting it to a vibratory force for determining qualities, characteristics, or conditions thereof, or sensing, studying or making analysis of the vibrations otherwise generated in or existing in the article or mechanism. Devices include methods to cause vibrations at a natural mechanical resonance and measure the frequency and/ or amplitude the resonance made. Various devices study the amplitude response over a frequency range is made. This includes nodal points, wave lengths, and standing wave characteristics measured under predetermined vibration conditions.

Earthquake Machine

Nikola Tesla established a laboratory at 46 E Houston Street in New York. There, at one point while experimenting with mechanical oscillators, he allegedly generated a resonance of several buildings causing complaints to the police. As the speed grew it is said that the machine oscillated at the resonance frequency of his own building and, belatedly realizing the danger, he was forced to use a sledge hammer to terminate the experiment, just as the police arrived. The Mythbusters television programme made a small machine based on the same principle, but driven by electricity rather than steam, to test the claimed earthquake effect; it produced vibrations in a large structure that could be felt hundreds of feet away, but no significant shaking, and they judged the effect to be a busted myth.

Degree of Freedom

In mechanics, the degree of freedom (DOF) of a mechanical system is the number of independent parameters that define its configuration. It is the number of parameters that determine the state of a physical system and is important to the analysis of systems of bodies in mechanical engineering, aeronautical engineering, robotics, and structural engineering.

The position of a single car (engine) moving along a track has one degree of freedom, because the position of the car is defined by the distance along the track. A train of rigid cars connected by hinges to an engine still has only one degree of freedom because the positions of the cars behind the engine are constrained by the shape of the track.

An automobile with highly stiff suspension can be considered to be a rigid body travelling on a plane (a flat, two-dimensional space). This body has three independent degrees of freedom consisting of two components of translation and one angle of rotation. Skidding or drifting is a good example of an automobile's three independent degrees of freedom. The position of a rigid body in space is defined by three components of translation and three components of rotation, which means that it has six degrees of freedom. The Exact constraint mechanical design method manages the degrees of freedom to neither underconstrain nor overconstrain a device.

Motions and Dimensions

The position of an n-dimensional rigid body is defined by the rigid transformation, [T]=[A, d], where d is an n-dimensional translation and

A is an n x n rotation matrix, which has n translational degrees of freedom and $n(n - 1)/2$ rotational degrees of freedom. The number of rotational degrees of freedom comes from the dimension of the rotation group SO(n).

A non-rigid or deformable body may be thought of as a collection of many minute particles (infinite number of DOFs); this is often approximated by a finite DOF system. When motion involving large displacements is the main objective of study (e.g. for analyzing the motion of satellites), a deformable body may be approximated as a rigid body (or even a particle) in order to simplify the analysis.

The degree of freedom of a system can be viewed as the minimum number of coordinates required to specify a configuration. Applying this definition, we have:

1. For a single particle in a plane two coordinates define its location so it has two degrees of freedom;
2. A single particle in space requires three coordinates so it has three degrees of freedom;
3. Two particles in space have a combined six degrees of freedom;
4. If two particles in space are constrained to maintain a constant distance from each other, such as in the case of a diatomic molecule, then the six coordinates must satisfy a single constraint equation defined by the distance formula. This reduces the degree of freedom of the system to five, because the distance formula can be used to solve for the remaining coordinate once the other five are specified.

Six Degrees of Freedom

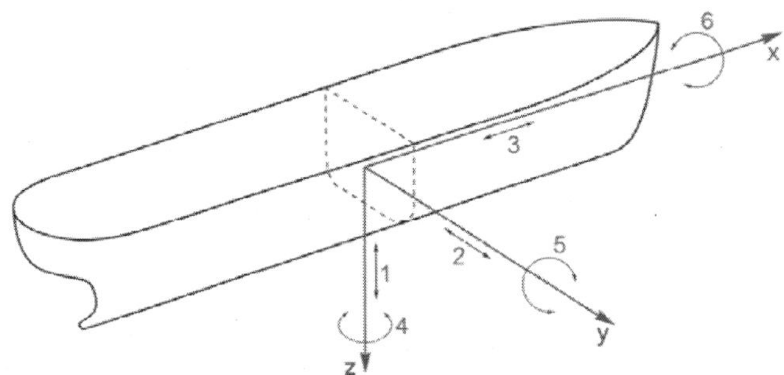

Figure: *The six degrees of freedom of movement of a ship.*

Figure: *Attitude degrees of freedom for an airplane.*

The motion of a ship at sea has the six degrees of freedom of a rigid body, and is described as:

Translation:

1. Moving up and down (heaving);
2. Moving left and right (swaying);
3. Moving forward and backward (surging);

Rotation

1. Tilts forward and backward (pitching);
2. Swivels left and right (yawing);
3. Pivots side to side (rolling).

The trajectory of an airplane in flight has three degrees of freedom and its attitude along the trajectory has three degrees of freedom, for a total of six degrees of freedom.

Mobility Formula

The mobility formula counts the number of parameters that define the configuration of a set of rigid bodies that are constrained by joints connecting these bodies.

Consider a system of n rigid bodies moving in space has $6n$ degrees of freedom measured relative to a fixed frame. In order to count the degrees of freedom of this system, include the ground frame in the count of bodies, so that mobility is independent of the choice of the body that forms the fixed frame. Then the degree-of-freedom of the unconstrained system of $N=n+1$ is

$$M = 6n = 6(N-1),$$

because the fixed body has zero degrees of freedom relative to itself.

Joints that connect bodies in this system remove degrees of freedom and reduce mobility. Specifically, hinges and sliders each impose five constraints and therefore remove five degrees of freedom. It is convenient to define the number of constraints c that a joint imposes in terms of the joint's freedom f, where $c=6-f$. In the case of a hinge or slider, which are one degree of freedom joints, have $f=1$ and therefore $c=6-1=5$.

The result is that the mobility of a system formed from n moving links and j joints each with freedom f_i, $i=1,..., j$, is given by

$$M = 6n - \sum_{i=1}^{j} (6 - f_i) = 6(N - 1 - j) + \sum_{i=1}^{j} f_i$$

Recall that N includes the fixed link.

There are two important special cases: (i) a simple open chain, and (ii) a simple closed chain. A single open chain consists of n moving links connected end to end by n joints, with one end connected to a ground link. Thus, in this case $N=j+1$ and the mobility of the chain is

$$M = \sum_{i=1}^{j} f_i$$

For a simple closed chain, n moving links are connected end-to-end by $n+1$ joints such that the two ends are connected to the ground link forming a loop. In this case, we have $N=j$ and the mobility of the chain is

$$M = \sum_{i=1}^{j} f_i - 6$$

An example of a simple open chain is a serial robot manipulator. These robotic systems are constructed from a series of links connected by six one degree-of-freedom revolute or prismatic joints, so the system has six degrees of freedom. An example of a simple closed chain is the RSSR spatial four-bar linkage. The sum of the freedom of these joints is eight, so the mobility of the linkage is two, where one of the degrees of freedom is the rotation of the coupler around the line joining the two S joints.

Planar and Spherical Movement

It is common practice to design the linkage system so that the movement of all of the bodies are constrained to lie on parallel planes, to form what is known as a *planar linkage*. It is also possible to construct the linkage system so that all of the bodies move on concentric spheres, forming a *spherical linkage*. In both cases, the degrees of freedom of the links in each system is now three rather than six, and the constraints imposed by joints are now $c=3-f$.

In this case, the mobility formula is given by

$$M = 3(N - 1 - j) + \sum_{i=1}^{j} f_i,$$

and the special cases become

- planar or spherical simple open chain,

$$M = \sum_{i=1}^{j} f_i,$$

- planar or spherical simple closed chain,

$$M = \sum_{i=1}^{j} f_i - 3.$$

An example of a planar simple closed chain is the planar four-bar linkage, which is a four-bar loop with four one degree-of-freedom joints and therefore has mobility M=1.

Systems of Bodies

A system with several bodies would have a combined DOF that is the sum of the DOFs of the bodies, less the internal constraints they may have on relative motion. A mechanism or linkage containing a number of connected rigid bodies may have more than the degrees of freedom for a single rigid body. Here the term *degrees of freedom* is used to describe the number of parameters needed to specify the spatial pose of a linkage.

A specific type of linkage is the open kinematic chain, where a set of rigid links are connected at joints; a joint may provide one DOF (hinge/sliding), or two (cylindrical). Such chains occur commonly in robotics, biomechanics, and for satellites and other space structures. A human arm is considered to have seven DOFs. A shoulder gives pitch, yaw, and roll, an elbow allows for pitch and roll, and a wrist allows for pitch and yaw. Only 3 of those movements would be necessary to move the hand to any point in space, but people would lack the ability to grasp things from different angles or directions. A robot (or object) that has mechanisms to control all 6 physical DOF is said to be holonomic. An object with fewer controllable DOFs than total DOFs is said to be non-holonomic, and an object with more controllable DOFs than total DOFs (such as the human arm) is said to be redundant.

In mobile robotics, a car-like robot can reach any position and orientation in 2-D space, so it needs 3 DOFs to describe its pose, but at any point, you can move it only by a forward motion and a steering angle. So it has two control DOFs and three representational DOFs; i.e. it is non-holonomic. A fixed-wing aircraft, with 3–4 control DOFs (forward motion, roll, pitch, and to a limited extent, yaw) in a 3-D space, is also non-holonomic, as it cannot move directly up/down or left/right.

A summary of formulas and methods for computing the degrees-of-freedom in mechanical systems has been given by Pennestri, Cavacece, and Vita.

Electrical Engineering

In electrical engineering *degrees of freedom* is often used to describe the number of directions in which a phased arrayantenna can form either beams or nulls. It is equal to one less than the number of elements contained in the array, as one element is used as a reference against which either constructive or destructive interference may be applied using each of the remaining antenna elements. radar practice and communication link practice, with beam steering being more prevalent for radar applications and null steering being more prevalent for interference suppression in communication links.

Types of Pendulum

A pendulum is a weight suspended from a pivot so that it can swing freely. When a pendulum is displaced sideways from its resting equilibrium position, it is subject to a restoring force due to gravity that will accelerate it back toward the equilibrium position. When released, the restoring force combined with the pendulum's mass causes it to oscillate about the equilibrium position, swinging back and forth. The time for one complete cycle, a left swing and a right swing, is called the period. A pendulum swings with a specific period which depends (mainly) on its length.

From its discovery around 1602 by Galileo Galilei the regular motion of pendulums was used for timekeeping, and was the world's most accurate timekeeping technology until the 1930s. Pendulums are used to regulate pendulum clocks, and are used in scientific instruments such as accelerometers and seismometers. Historically they were used as gravimeters to measure the acceleration of gravity in geophysical surveys, and even as a standard of length. The word 'pendulum' is new Latin, from the Latin *pendulus*, meaning 'hanging'.

The simple gravity pendulum is an idealized mathematical model of a pendulum. This is a weight (or bob) on the end of a massless cord suspended from a pivot, without friction. When given an initial push, it will swing back and forth at a constant amplitude. Real pendulums are subject to friction and air drag, so the amplitude of their swings declines.

Period of Oscillation

The period of swing of a simple gravity pendulum depends on its length, the local strength of gravity, and to a small extent on the maximum angle that the pendulum swings away from vertical, θ_0, called the amplitude. It is independent of the mass of the bob. If the amplitude is limited to small swings, the period T of a simple pendulum, the time taken for a complete cycle, is:

$$T \approx 2\pi\sqrt{\frac{L}{g}} \qquad \theta_0 \ll 1 \qquad (1)$$

where L is the length of the pendulum and g is the local acceleration of gravity.

For small swings the period of swing is approximately the same for different size swings: that is, *the period is independent of amplitude.* This property, called isochronism, is the reason pendulums are so useful for timekeeping. Successive swings of the pendulum, even if changing in amplitude, take the same amount of time.

For larger amplitudes, the period increases gradually with amplitude so it is longer than given by equation. For example, at an amplitude of θ_0 = 23° it is 1% larger than given by (1). The period increases asymptotically (to infinity) as θ_0 approaches 180°, because the value θ_0 = 180° is an unstable equilibrium point for the pendulum. The true period of an ideal simple gravity pendulum can be written in several different forms, one example being the infinite series:

$$T = 2\pi\sqrt{\frac{L}{g}}\left(1 + \frac{1}{16}\theta_0^2 + \frac{11}{3072}\theta_0^4 + \cdots\right)$$

The difference between this true period and the period for small swings (1) above is called the *circular error.* In the case of a longcase clock whose pendulum is about one metre in length and whose amplitude is ±0.1 radians, the θ^2 term adds a correction to equation (1) that is equivalent to 54 seconds per day and the θ^4 term a correction equivalent to a further 0.03 seconds per day.

For small swings the pendulum approximates a harmonic oscillator, and its motion as a function of time, t, is approximately simple harmonic motion:

$$\theta(t) = \theta_0 \cos(2\pi t / T).$$

For real pendulums, corrections to the period may be needed to take into account the presence of air, the mass of the string, the size and shape of the bob and how it is attached to the string, flexibility and stretching of the string, motion of the support, and local gravitational gradients.

Compound Pendulum

The length L of the ideal simple pendulum above, used for calculating the period, is the distance from the pivot point to the centre of mass of the bob. A pendulum consisting of any swinging rigid body, which is free to rotate about a fixed horizontal axis is called a compound pendulum or physical pendulum. For these pendulums the appropriate equivalent length is the distance from the pivot point to a point in the pendulum called the *centre of oscillation.* This is located under the centre of mass, at a distance called the radius of gyration, that depends on the mass distribution along the pendulum. However, for any pendulum in which most of the mass is concentrated in the bob, the centre of oscillation is close to the centre of mass.

Using the parallel axis theorem, the radius of gyration L of a rigid pendulum can be shown to be

$$L = \frac{I}{mR}.$$

Substituting this into (1) above, the period T of a rigid-body compound pendulum for small angles is given by

$$T = 2\pi\sqrt{\frac{I}{mgR}},$$

where$I$ is the moment of inertia of the pendulum about the pivot point, m is the mass of the pendulum, and R is the distance between the pivot point and the centre of mass of the pendulum.

For example, for a pendulum made of a rigid uniform rod of length L pivoted at its end, $I = (1/3)mL^2$. The centre of mass is located in the centre of the rod, so $R = L/2$. Substituting these values into the above equation gives $T = 2\pi\sqrt{2L/3g}$. This shows that a rigid rod pendulum has the same period as a simple pendulum of 2/3 its length.

Christiaan Huygens proved in 1673 that the pivot point and the centre of oscillation are interchangeable. This means if any pendulum is turned upside down and swung from a pivot located at its previous centre of oscillation, it will have the same period as before, and the new centre of oscillation will be at the old pivot point. In 1817 Henry Kater used this idea to produce a type of reversible pendulum, now known as a Kater pendulum, for improved measurements of the acceleration due to gravity.

History

One of the earliest known uses of a pendulum was in the 1st. century seismometer device of Han Dynasty Chinese scientist Zhang

Heng. Its function was to sway and activate one of a series of levers after being disturbed by the tremor of an earthquake far away. Released by a lever, a small ball would fall out of the urn-shaped device into one of eight metal toad's mouths below, at the eight points of the compass, signifying the direction the earthquake was located.

Many sources claim that the 10th century Egyptian astronomer IbnYunus used a pendulum for time measurement, but this was an error that originated in 1684 with the British historian Edward Bernard.

During the Renaissance, large pendulums were used as sources of power for manual reciprocating machines such as saws, bellows, and pumps.Leonardo da Vinci made many drawings of the motion of pendulums, though without realizing its value for timekeeping.

1602: Galileo's Research

Italian scientist Galileo Galilei was the first to study the properties of pendulums, beginning around 1602. His first existent report of his research is contained in a letter to Guido Ubaldo dal Monte, from Padua, dated November 29, 1602. His biographer and student, Vincenzo Viviani, claimed his interest had been sparked around 1582 by the swinging motion of a chandelier in the Pisa cathedral. Galileo discovered the crucial property that makes pendulums useful as timekeepers, called isochronism; the period of the pendulum is approximately independent of the amplitude or width of the swing. He also found that the period is independent of the mass of the bob, and proportional to the square root of the length of the pendulum. He first employed freeswinging pendulums in simple timing applications. A physician friend invented a device which measured a patient's pulse by the length of a pendulum; the *pulsilogium*. In 1641 Galileo conceived and dictated to his son Vincenzo a design for a pendulum clock; Vincenzo began construction, but had not completed it when he died in 1649. The pendulum was the first harmonic oscillator used by man.

1656: The Pendulum Clock

In 1656 the Dutch scientist Christiaan Huygens built the first pendulum clock. This was a great improvement over existing mechanical clocks; their best accuracy was increased from around 15 minutes deviation a day to around 15 seconds a day. Pendulums spread over Europe as existing clocks were retrofitted with them.

The English scientist Robert Hooke studied the conical pendulum around 1666, consisting of a pendulum that is free to swing in two dimensions, with the bob rotating in a circle or ellipse. He used the motions of this device as a model to analyze the orbital motions of the planets. Hooke suggested to Isaac Newton in 1679 that the components of orbital motion consisted of inertial motion along a tangent direction plus an attractive motion in the radial direction. This played a part in Newton's formulation of the law of universal gravitation. Robert Hooke was also responsible for suggesting as early as 1666 that the pendulum could be used to measure the force of gravity.

During his expedition to Cayenne, French Guiana in 1671, Jean Richer found that a pendulum clock was 2 $^{1}/_{2}$ minutes per day slower at Cayenne than at Paris. From this he deduced that the force of gravity was lower at Cayenne. In 1687, Isaac Newton in *Principia Mathematica* showed that this was because the Earth was not a true sphere but slightly oblate (flattened at the poles) from the effect of centrifugal force due to its rotation, causing gravity to increase with latitude. Portable pendulums began to be taken on voyages to distant lands, as precision gravimeters to measure the acceleration of gravity at different points on Earth, eventually resulting in accurate models of the shape of the Earth.

1673: Huygens' Horologium Oscillatorium

In 1673, Christiaan Huygens published his theory of the pendulum, *Horologium Oscillatoriumsive de motupendulorum*. He demonstrated that for an object to descend down a curve under gravity in the same time interval, regardless of the starting point, it must follow a cycloid curve rather than the circular arc of a pendulum. This confirmed the earlier observation by Marin Mersenne that the period of a pendulum does vary with its amplitude, and that Galileo's observation of isochronism was accurate only for small swings. Huygens also solved the issue of how to calculate the period of an arbitrarily shaped pendulum (called a *compound pendulum*), discovering the *centre of oscillation*, and its interchangeability with the pivot point.

The existing clock movement, the verge escapement, made pendulums swing in very wide arcs of about 100°. Huygens showed this was a source of inaccuracy, causing the period to vary with amplitude changes caused by small unavoidable variations in the clock's drive force. To make its period isochronous, Huygens mounted cycloidal-shaped metal 'cheeks' next to the pivot in his 1673 clock, that constrained the suspension cord and forced the pendulum to follow a cycloid arc. This solution didn't prove as practical as simply limiting the pendulum's swing to small angles of a few degrees. The realization that only small swings were isochronous motivated the development of the anchor escapement around 1670, which reduced the pendulum swing in clocks to 4°–6°.

1721: Temperature Compensated Pendulums

During the 18th and 19th century, the pendulum clock's role as the most accurate timekeeper motivated much practical research into improving pendulums. It was found that a major source of error was that the pendulum rod expanded and contracted with changes in ambient temperature, changing the period of swing. This was solved with the invention of temperature compensated pendulums, the mercury pendulum in 1721 and the gridiron pendulum in 1726, reducing errors in precision pendulum clocks to a few seconds per week.

The accuracy of gravity measurements made with pendulums was limited by the difficulty of finding the location of their centre of oscillation. Huygens had discovered in 1673 that a pendulum has the same period when hung from its centre of oscillation as when hung from its pivot, and the distance between the two points was equal to the length of a simple gravity pendulum of the same period. In 1818 British Captain Henry Kater invented the reversible Kater'spendulum which used this principle, making possible very accurate measurements of gravity. For the next century the reversible pendulum was the standard method of measuring absolute gravitational acceleration.

1851: Foucault Pendulum

In 1851, Jean Bernard Léon Foucault showed that the plane of oscillation of a pendulum, like a gyroscope, tends to stay constant regardless of the motion of the pivot, and that this could be used to demonstrate the rotation of the Earth. He suspended a pendulum free to swing in two dimensions (later named the Foucault pendulum) from the dome of the Panthéon in Paris. The length of the cord was 67 m (220 ft). Once the pendulum was set in motion, the plane of swing was observed to precess or rotate 360° clockwise in about 32 hours. This was the first demonstration of the Earth's rotation that didn't depend on celestial observations, and a "pendulum mania" broke out, as Foucault pendulums were displayed in many cities and attracted large crowds.

1930: Decline in Use

Around 1900 low-thermal-expansion materials began to be used for pendulum rods in the highest precision clocks and other instruments, first invar, a nickel steel alloy, and later fused quartz, which made temperature compensation trivial. Precision pendulums were housed in low pressure tanks, which kept the air pressure constant to prevent changes in the period due to changes in buoyancy of the pendulum due to changing atmospheric pressure. The accuracy of the best pendulum clocks topped out at around a second per year.

The timekeeping accuracy of the pendulum was exceeded by the quartzcrystal oscillator, invented in 1921, and quartz clocks, invented in 1927, replaced pendulum clocks as the world's best timekeepers. Pendulum clocks were used as time standards until World War 2, although the French Time Service continued using them in their official time standard ensemble until 1954. Pendulum gravimeters were superseded by "free fall" gravimeters in the 1950s, but pendulum instruments continued to be used into the 1970s.

Use for Time Measurement

For 300 years, from its discovery around 1602 until development of the quartz clock in the 1930s, the pendulum was the world's standard for accurate timekeeping. In addition to clock pendulums, free swinging seconds pendulums were widely used as precision timers in scientific experiments in the 17th and 18th centuries. Pendulums require great mechanical stability: a length change of only 0.02%, 0.2 mm in a grandfather clock pendulum, will cause an error of a minute per week.

Clock Pendulums

Pendulums in clocks are usually made of a weight or bob*(b)* suspended by a rod of wood or metal *(a)*. To reduce air resistance (which accounts for most of the energy loss in clocks) the bob is traditionally a smooth disk with a lens-shaped cross section, although in antique clocks it often had carvings or decorations specific to the type of clock. In quality clocks the bob is made as heavy as the suspension can support and the movement can drive, since this improves the regulation of the clock. A common weight for seconds pendulum bobs is 15 pounds. (6.8 kg). Instead of hanging from a pivot, clock pendulums are usually supported by a short straight spring*(d)* of flexible metal ribbon. This avoids the friction and 'play' caused by a pivot, and the slight bending force of the spring merely adds to the pendulum's restoring force. A few precision clocks have pivots of 'knife' blades resting on agate plates. The impulses to keep the pendulum swinging are provided by an arm hanging behind the pendulum called the *crutch*, *(e)*, which ends in a *fork*, *(f)* whose prongs embrace the pendulum rod. The crutch is pushed back and forth by the clock's escapement, *(g,h)*.

Each time the pendulum swings through its centre position, it releases one tooth of the *escape wheel(g)*. The force of the clock's mainspring or a driving weight hanging from a pulley, transmitted through the clock's gear train, causes the wheel to turn, and a tooth presses against one of the pallets *(h)*, giving the pendulum a short push. The clock's wheels, geared to the escape wheel, move forward a fixed amount with each pendulum swing, advancing the clock's hands at a steady rate.

The pendulum always has a means of adjusting the period, usually by an adjustment nut *(c)* under the bob which moves it up or down on the rod. Moving the bob up decreases the pendulum's length, causing the pendulum to swing faster and the clock to gain time. Some precision clocks have a small auxiliary adjustment weight on a threaded shaft on the bob, to allow finer adjustment. Some tower clocks and precision clocks use a tray attached near to the midpoint of the pendulum rod, to which small weights can be added or removed. This effectively shifts the centre of oscillation and allows the rate to be adjusted without stopping the clock.

The pendulum must be suspended from a rigid support. During operation, any elasticity will allow tiny imperceptible swaying motions of the support, which disturbs the clock's period, resulting in error. Pendulum clocks should be attached firmly to a sturdy wall. The most common pendulum length in quality clocks, which is always used in grandfather clocks, is the seconds pendulum, about 1 metre (39 inches) long. In mantel clocks, half-second pendulums, 25 cm (10 in) long, or shorter, are used. Only a few large tower clocks use longer pendulums, the 1.5 second pendulum, 2.25 m (7 ft) long, or occasionally the two-second pendulum, 4 m (13 ft) as is the case of Big Ben.

Temperature Compensation

The largest source of error in early pendulums was slight changes in length due to thermal expansion and contraction of the pendulum rod with changes in ambient temperature. This was discovered when people noticed that pendulum clocks ran slower in summer, by as much as a minute per week (one of the first was Godefroy Wendelin, as reported by Huygens in 1658). Thermal expansion of pendulum rods was first studied by Jean Picard in 1669. A pendulum with a steel rod will expand by about 11.3 parts per million (ppm) with each degree Celsius increase, causing it to lose about 0.27 seconds per day for every degree Celsius increase in temperature, or 9 seconds per day for a 33 °C (60 °F) change. Wood rods expand less, losing only about 6 seconds per day for a 33 °C (60 °F) change, which is why quality clocks often had wooden pendulum rods. However, care had to be taken to reduce the possibility of errors due to changes in humidity.

Mercury Pendulum

The first device to compensate for this error was the mercury pendulum, invented by George Graham in 1721. The liquid metal mercury expands in volume with temperature. In a mercury pendulum, the pendulum's weight (bob) is a container of mercury. With a temperature rise, the pendulum rod gets longer, but the mercury also expands and its surface level rises slightly in the container, moving its centre of mass closer to the pendulum pivot. By using the correct height of mercury in the container these two effects will cancel, leaving the pendulum's centre of mass, and its period, unchanged with temperature. Its main disadvantage was that when the temperature changed, the rod would come to the new temperature quickly but the mass of mercury might take a day or two to reach the new temperature, causing the rate to deviate during that time. To improve thermal accommodation several thin containers were often used, made of metal. Mercury pendulums were the standard used in precision regulator clocks into the 20th century.

Gridiron Pendulum

The most widely used compensated pendulum was the gridiron pendulum, invented in 1726 by John Harrison. This consists of alternating rods of two different metals, one with lower thermal expansion (CTE), steel, and one with higher thermal expansion, zinc or brass. The rods are connected by a frame, so that an increase in length of the zinc rods pushes the bob up, shortening the pendulum. With a temperature increase, the low expansion steel rods make the pendulum longer, while the high expansion zinc rods make it shorter. By making the rods of the correct lengths, the greater expansion of the zinc cancels out the expansion of the steel rods which have a greater combined length, and the pendulum stays the same length with temperature.

Zinc-steel gridiron pendulums are made with 5 rods, but the thermal expansion of brass is closer to steel, so brass-steel gridirons usually require 9 rods. Gridiron pendulums adjust to temperature changes faster than mercury pendulums, but scientists found that friction of the rods sliding in their holes in the frame caused gridiron pendulums to adjust in a series of tiny jumps. In high precision clocks this caused the clock's rate to change suddenly with each jump. Later it was found that zinc is subject to creep. For these reasons mercury pendulums were used in the highest precision clocks, but gridirons were used in quality regulator clocks. They became so associated with quality that, to this day, many ordinary clock pendulums have decorative 'fake' gridirons that don't actually have any temperature compensation function.

Invar and Fused Quartz

Around 1900 low thermal expansion materials were developed which, when used as pendulum rods, made elaborate temperature compensation unnecessary. These were only used in a few of the highest precision clocks before the pendulum became obsolete as a time standard. In 1896 Charles Edouard Guillaume invented the nickelsteelalloyInvar. This has a CTE of around 0.5 µin/(in·°F), resulting in pendulum temperature errors over 71 °F of only 1.3 seconds per day, and this residual error could be compensated to zero with a few centimeters of aluminium under the pendulum bob (this can be seen in the Riefler clock image above). Invar pendulums were first used in 1898 in the Riefler regulator clock which achieved accuracy of 15 milliseconds per day. Suspension springs of Elinvar were used to eliminate temperature variation of the spring's restoring force on the pendulum. Later fused quartz was used which had even lower CTE. These materials are the choice for modern high accuracy pendulums.

Atmospheric Pressure

The effect of the surrounding air on a moving pendulum is complex and requires fluid mechanics to calculate precisely, but for most purposes its influence on the period can be accounted for by three effects:

- By Archimedes' principle the effective weight of the bob is reduced by the buoyancy of the air it displaces, while the mass (inertia) remains the same, reducing the pendulum's acceleration during its swing and increasing the period. This depends on the air pressure and the density of the pendulum, but not its shape.
- The pendulum carries an amount of air with it as it swings, and the mass of this air increases the inertia of the pendulum, again reducing the acceleration and increasing the period. This depends on both its density and shape.
- Viscous air resistance slows the pendulum's velocity. This has a negligible effect on the period, but dissipates energy, reducing the amplitude. This reduces the pendulum's Q factor, requiring a stronger drive force from the clock's mechanism to keep it moving, which causes increased disturbance to the period.

So increases in barometric pressure increase a pendulum's period slightly due to the first two effects, by about 0.11 seconds per day per kilopascal (0.37 seconds per day per inch of mercury or 0.015 seconds per day per torr). Researchers using pendulums to measure the acceleration of gravity had to correct the period for the air pressure at the altitude of measurement, computing the equivalent period of a pendulum swinging in vacuum. A pendulum clock was first operated in a constant-pressure tank by Friedrich Tiede in 1865 at the Berlin Observatory, and by 1900 the highest precision clocks were mounted in tanks that were kept at a constant pressure to eliminate changes in atmospheric pressure. Alternatively, in some a small aneroid barometer mechanism attached to the pendulum compensated for this effect.

Gravity

Pendulums are affected by changes in gravitational acceleration, which varies by as much as 0.5% at different locations on Earth, so pendulum clocks have to be recalibrated after a move. Even moving a pendulum clock to the top of a tall building can cause it to lose measurable time from the reduction in gravity.

Accuracy of Pendulums as Timekeepers

The timekeeping elements in all clocks, which include pendulums, balance wheels, the quartz crystals used in quartz watches, and even the vibrating atoms in atomic clocks, are in physics called harmonic oscillators. The reason harmonic oscillators are used in clocks is that they vibrate or oscillate at a specific resonant frequency or period and resist oscillating at other rates. However, the resonant frequency is not infinitely 'sharp'. Around the resonant frequency there is a narrow natural band of frequencies (or periods), called the resonance width or bandwidth, where the harmonic oscillator will oscillate. In a clock, the actual frequency of the pendulum may vary randomly within this bandwidth in response to disturbances, but at frequencies outside this band, the clock will not function at all.

Q Factor

The measure of a harmonic oscillator's resistance to disturbances to its oscillation period is a dimensionless parameter called the Q factor equal to the resonant frequency divided by the bandwidth. The higher the Q, the smaller the bandwidth, and the more constant the frequency or period of the oscillator for a given disturbance. The reciprocal of the Q is roughly proportional to the limiting accuracy achievable by a harmonic oscillator as a time standard.

The Q is related to how long it takes for the oscillations of an oscillator to die out. The Q of a pendulum can be measured by counting the number of oscillations it takes for the amplitude of the pendulum's swing to decay to $1/e$ = 36.8% of its initial swing, and multiplying by 2π.

In a clock, the pendulum must receive pushes from the clock's movement to keep it swinging, to replace the energy the pendulum loses to friction. These pushes, applied by a mechanism called the escapement, are the main source of disturbance to the pendulum's motion. The Q is equal to 2π times the energy stored in the pendulum, divided by the energy lost to friction during each oscillation period, which is the same as the energy added by the escapement each period. It can be seen that the smaller the fraction of the pendulum's energy that is lost to friction, the less energy needs to be added, the less the disturbance from the escapement, the more 'independent' the pendulum is of the clock's mechanism, and the more constant its period is. The Q of a pendulum is given by:

$$Q = \frac{M\omega}{\Gamma}$$

where$M$ is the mass of the bob, $\omega = 2\pi/T$ is the pendulum's radian frequency of oscillation, and Γ is the frictional damping force on the pendulum per unit velocity.

ω is fixed by the pendulum's period, and M is limited by the load capacity and rigidity of the suspension. So the Q of clock pendulums is increased by minimizing frictional losses (Γ). Precision pendulums are suspended on low friction pivots consisting of triangular shaped 'knife' edges resting on agate plates. Around 99% of the energy loss in a freeswinging pendulum is due to air friction, so mounting a pendulum in a vacuum tank can increase the Q, and thus the accuracy, by a factor of 100.

The Q of pendulums ranges from several thousand in an ordinary clock to several hundred thousand for precision regulator pendulums swinging in vacuum. A quality home pendulum clock might have a Q of 10,000 and an accuracy of 10 seconds per month. The most accurate commercially produced pendulum clock was the Shortt-Synchronome free pendulum clock, invented in 1921. Its Invar master pendulum swinging in a vacuum tank had a Q of 110,000 and an error rate of around a second per year.

Their Q of 10^3–10^5 is one reason why pendulums are more accurate timekeepers than the balance wheels in watches, with Q around 100-300, but less accurate than the quartz crystals in quartz clocks, with Q of 10^5–10^6.

Escapement

Pendulums (unlike, for example, quartz crystals) have a low enough Q that the disturbance caused by the impulses to keep them moving is generally the limiting factor on their timekeeping accuracy. Therefore the design of the escapement, the mechanism that provides these impulses, has a large effect on the accuracy of a clock pendulum. If the impulses given to the pendulum by the escapement each swing could be exactly identical, the response of the pendulum would be identical, and its period would be constant. However, this is not achievable; unavoidable random fluctuations in the force due to friction of the clock's pallets, lubrication variations, and changes in the torque provided by the clock's power source as it runs down, mean that the force of the impulse applied by the escapement varies.

If these variations in the escapement's force cause changes in the pendulum's width of swing (amplitude), this will cause corresponding slight changes in the period, since (as discussed at top) a pendulum with a finite swing is not quite isochronous. Therefore, the goal of traditional escapement design is to apply the force with the proper profile, and at the correct point in the pendulum's cycle, so force variations have no effect on the pendulum's amplitude. This is called an *isochronous escapement.*

The Airy Condition

In 1826 British astronomer George Airy proved what clockmakers had known for centuries; that the disturbing effect of a drive force on the period of a pendulum is smallest if given as a short impulse as the pendulum passes through its bottom equilibrium position. Specifically, he proved that if a pendulum is driven by an impulse that is symmetrical about its bottom equilibrium position, the pendulum's amplitude will be unaffected by changes in the drive force. The most accurate escapements, such as the deadbeat, approximately satisfy this condition.

Gravity Measurement

The presence of the acceleration of gravity$g$ in the periodicity equation (1) for a pendulum means that the local gravitational acceleration of the Earth can be calculated from the period of a pendulum. A pendulum can therefore be used as a gravimeter to measure the local gravity, which varies by over 0.5% across the surface of the Earth. The pendulum in a clock is disturbed by the pushes it receives from the clock movement, so freeswinging pendulums were used, and were the standard instruments of gravimetry up to the 1930s.

The difference between clock pendulums and gravimeter pendulums is that to measure gravity, the pendulum's length as well as its period has to be measured. The period of freeswinging pendulums could be found to great precision by comparing their swing with a precision clock that had been adjusted to keep correct time by the passage of stars overhead. In the early measurements, a weight on a cord was suspended in front of the clock pendulum, and its length adjusted until the two pendulums swung in exact synchronism. Then the length of the cord was measured. From the length and the period, g could be calculated from (1).

The Seconds Pendulum

The seconds pendulum, a pendulum with a period of two seconds so each swing takes one second, was widely used to measure gravity, because most precision clocks had seconds pendulums. By the late 17th century, the length of the seconds pendulum became the standard measure of the strength of gravitational acceleration at a location. By 1700 its length had been measured with submillimeter accuracy at several cities in Europe. For a seconds pendulum, g is proportional to its length:

$$g \propto L.$$

Kater's Pendulum

The precision of the early gravity measurements above was limited by the difficulty of measuring the length of the pendulum, L .L was the length of an idealized simple gravity pendulum (described at top), which has all its mass concentrated in a point at the end of the cord. In 1673 Huygens had shown that the period of a real pendulum (called a *compound pendulum*) was equal to the period of a simple pendulum with a length equal to the distance between the pivot point and a point called the centre of oscillation, located under the centre of gravity, that depends on the mass distribution along the pendulum. But there was no accurate way of determining the centre of oscillation in a real pendulum.

To get around this problem, the early researchers above approximated an ideal simple pendulum as closely as possible by using a metal sphere suspended by a light wire or cord. If the wire was light enough, the centre of oscillation was close to the centre of gravity of the ball, at its geometric centre. This "ball and wire" type of pendulum wasn't very accurate, because it didn't swing as a rigid body, and the elasticity of the wire caused its length to change slightly as the pendulum swung.

However Huygens had also proved that in any pendulum, the pivot point and the centre of oscillation were interchangeable. That is, if a pendulum were turned upside down and hung from its centre of oscillation, it would have the same period as it did in the previous position, and the old pivot point would be the new centre of oscillation.

British physicist and army captain Henry Kater in 1817 realized that Huygens' principle could be used to find the length of a simple pendulum with the same period as a real pendulum. If a pendulum was built with a second adjustable pivot point near the bottom so it could be hung upside down, and the second pivot was adjusted until the periods when hung from both pivots were the same, the second pivot would be at the centre of oscillation, and the distance between the two pivots would be the length of a simple pendulum with the same period.

Kater built a reversible pendulum (shown at right) consisting of a brass bar with two opposing pivots made of short triangular "knife" blades *(a)* near either end. It could be swung from either pivot, with the knife blades supported on agate plates. Rather than make one pivot adjustable, he attached the pivots a metre apart and instead adjusted the periods with a moveable weight on the pendulum rod *(b,c)*. In operation, the pendulum is hung in front of a precision clock, and the period timed, then turned upside down and the period timed again. The weight is adjusted with the adjustment screw until the periods are equal. Then putting this period and the distance between the pivots into equation (1) gives the gravitational acceleration *g* very accurately.

Kater timed the swing of his pendulum using the "*method of coincidences*" and measured the distance between the two pivots with a microscope. After applying corrections for the finite amplitude of swing, the buoyancy of the bob, the barometric pressure and altitude, and temperature, he obtained a value of 39.13929 inches for the seconds pendulum at London, in vacuum, at sea level, at 62 °F. The largest variation from the mean of his 12 observations was 0.00028 in. representing a precision of gravity measurement of 7×10^{-6} (7 mGal or 70 $\mu m/s^2$). Kater's measurement was used as Britain's official standard of length from 1824 to 1855.

Reversible pendulums (known technically as "convertible" pendulums) employing Kater's principle were used for absolute gravity measurements into the 1930s.

Later Pendulum Gravimeters

The increased accuracy made possible by Kater's pendulum helped make gravimetry a standard part of geodesy. Since the exact location (latitude and longitude) of the 'station' where the gravity measurement was made was necessary, gravity measurements became part of surveying, and pendulums were taken on the great geodetic surveys of the 18th century, particularly the Great Trigonometric Survey of India.

- *Invariable Pendulums:* Kater introduced the idea of *relative* gravity measurements, to supplement the *absolute* measurements made by a Kater's pendulum. Comparing the gravity at two different points was an easier process than measuring it absolutely by the Kater method. All that was necessary was to time the period of an ordinary (single pivot) pendulum at the first point, then transport the pendulum to the other point and time its period there. Since the pendulum's length was constant, from (1) the ratio of the gravitational accelerations was equal to the inverse of the ratio of the periods squared, and no precision length measurements were necessary. So once the gravity had been measured absolutely at some central station, by the Kater or other accurate method, the gravity at other points could be found by swinging pendulums at the central station and then taking them to the nearby point. Kater made up a set of "invariable" pendulums, with only one knife edge pivot, which were taken to many countries after first being swung at a central station at Kew Observatory, UK.
- *Airy's Coal Pit Experiments:* Starting in 1826, using methods similar to Bouguer, British astronomer George Airy attempted to determine the density of the Earth by pendulum gravity measurements at the top and bottom of a coal mine. The gravitational force below the surface of the Earth decreases rather than increasing with depth, because by Gauss's law the mass of the spherical shell of crust above the subsurface point does not contribute to the gravity. The 1826 experiment was aborted by the flooding of the mine, but in 1854 he conducted an improved experiment at the Harton coal mine, using seconds pendulums swinging on agate plates, timed by precision chronometers synchronized by an electrical circuit. He found the lower pendulum was slower by 2.24 seconds per day. This meant that the gravitational acceleration at the bottom of the mine, 1250 ft below the surface, was 1/14,000 less than it should have been from the inverse square law; that is the attraction of the spherical shell was 1/14,000 of the attraction of the Earth. From samples of surface rock he estimated the mass of the

spherical shell of crust, and from this estimated that the density of the Earth was 6.565 times that of water. Von Sterneck attempted to repeat the experiment in 1882 but found inconsistent results.

- *Repsold-Bessel Pendulum:* It was time-consuming and error-prone to repeatedly swing the Kater's pendulum and adjust the weights until the periods were equal. Friedrich Bessel showed in 1835 that this was unnecessary. As long as the periods were close together, the gravity could be calculated from the two periods and the centre of gravity of the pendulum.So the reversible pendulum didn't need to be adjustable, it could just be a bar with two pivots. Bessel also showed that if the pendulum was made symmetrical in form about its centre, but was weighted internally at one end, the errors due to air drag would cancel out. Further, another error due to the finite diameter of the knife edges could be made to cancel out if they were interchanged between measurements. Bessel didn't construct such a pendulum, but in 1864 Adolf Repsold, under contract by the Swiss Geodetic Commission made a pendulum along these lines. The Repsold pendulum was about 56 cm long and had a period of about $^3/_4$ second. It was used extensively by European geodetic agencies, and with the Kater pendulum in the Survey of India. Similar pendulums of this type were designed by Charles Pierce and C. Defforges.
- *Von Sterneck and Mendenhall Gravimeters:* In 1887 Austro-Hungarian scientist Robert von Sterneck developed a small gravimeter pendulum mounted in a temperature-controlled vacuum tank to eliminate the effects of temperature and air pressure. These used "half-second pendulums," having a period close to one second, and were about 25 cm long. They were nonreversible, so it was used for relative gravity measurements, but their small size made them small and portable. The period of the pendulum was picked off by reflecting the image of an electric spark created by a precision chronometer off a mirror mounted at the top of the pendulum rod. The Von Sterneck instrument, and a similar instrument developed by Thomas C. Mendenhall of the US Coast and Geodetic Survey in 1890, were used extensively for surveys into the 1920s.

The Mendenhall pendulum was actually a more accurate timekeeper than the highest precision clocks of the time, and as the 'world's best clock' it was used by A. A. Michelson in his 1924 measurements of the speed of light on Mt. Wilson, California.

- Double pendulum gravimeters: Starting in 1875, the increasing accuracy of pendulum measurements revealed another source of error in existing instruments: the swing of the pendulum caused a slight swaying of the tripod stand used to support portable pendulums, introducing error. In 1875 Charles S Peirce calculated that measurements of the length of the seconds pendulum made with the Repsold instrument required a correction of 0.2 mm due to this error. In 1880 C. Defforges used a Michelson interferometer to measure the sway of the stand dynamically, and interferometers were added to the standard Mendenhall apparatus to calculate sway corrections. A method of preventing this error was first suggested in 1877 by Hervé Faye and advocated by Peirce, Cellérier and Furtwangler: mount two identical pendulums on the same support, swinging with the same amplitude, 180° out of phase. The opposite motion of the pendulums would cancel out any sideways forces on the support. The idea was opposed due to its complexity, but by the start of the 20th century the Von Sterneck device and other instruments were modified to swing multiple pendulums simultaneously.
- *Gulf gravimeter:* One of the last and most accurate pendulum gravimeters was the apparatus developed in 1929 by the Gulf Research and Development Co. It used two pendulums made of fused quartz, each 10.7 inches (272 mm) in length with a period of 0.89 second, swinging on pyrex knife edge pivots, 180° out of phase. They were mounted in a permanently sealed temperature and humidity controlled vacuum chamber. Stray electrostatic charges on the quartz pendulums had to be discharged by exposing them to a radioactive salt before use. The period was detected by reflecting a light beam from a mirror at the top of the pendulum, recorded by a chart recorder and compared to a precision crystal oscillator calibrated against the WWV radio time signal. This instrument was accurate to within $(0.3–0.5)\times10^{-7}$ (30–50 microgals or 3–5 nm/s^2).It was used into the 1960s.

Relative pendulum gravimeters were superseded by the simpler LaCoste zero-length spring gravimeter, invented in 1934 by Lucien LaCoste. Absolute (reversible) pendulum gravimeters were replaced in the 1950s by free fall gravimeters, in which a weight is allowed to fall in a vacuum tank and its acceleration is measured by an optical interferometer.

Standard of Length

Because the acceleration of gravity is constant at a given point on Earth, the period of a simple pendulum at a given location depends only on its length. Additionally, gravity varies only slightly at different locations. Almost from the pendulum's discovery until the early 19th century, this property led scientists to suggest using a pendulum of a given period as a standard of length.

Until the 19th century, countries based their systems of length measurement on prototypes, metal bar primary standards, such as the standard yard in Britain kept at the Houses of Parliament, and the standard *toise* in France, kept at Paris. These were vulnerable to damage or destruction over the years, and because of the difficulty of comparing prototypes, the same unit often had different lengths in distant towns, creating opportunities for fraud.Enlightenment scientists argued for a length standard that was based on some property of nature that could be determined by measurement, creating an indestructible, universal standard. The period of pendulums could be measured very precisely by timing them with clocks that were set by the stars. A pendulum standard amounted to defining the unit of length by the gravitational force of the Earth, for all intents constant, and the second, which was defined by the rotation rate of the Earth, also constant. The idea was that anyone, anywhere on Earth, could recreate the standard by constructing a pendulum that swung with the defined period and measuring its length.

Virtually all proposals were based on the seconds pendulum, in which each swing (a half period) takes one second, which is about a metre (39 inches) long, because by the late 17th century it had become a standard for measuring gravity. By the 18th century its length had been measured with sub-millimeter accuracy at a number of cities in Europe and around the world.

The initial attraction of the pendulum length standard was that it was believed (by early scientists such as Huygens and Wren) that gravity was constant over the Earth's surface, so a given pendulum had the same period at any point on Earth.So the length of the standard pendulum could be measured at any location, and would not be tied to any given nation or region; it would be a truly democratic, worldwide standard. Although Richer found in 1672 that gravity varies at different points on the globe, the idea of a pendulum length standard remained popular, because it was found that gravity only varies with latitude. Gravitational acceleration increases smoothly from the equator to the poles, due to the oblate shape of the Earth. So at any given latitude (east-west line), gravity was constant enough that the length of a seconds pendulum was the same within the measurement capability of the 18th century. So the unit of length could be defined at a given latitude and measured at any point at that latitude. For example, a pendulum standard defined at 45° north latitude, a popular choice, could be measured in parts of France, Italy, Croatia, Serbia, Romania, Russia, Kazakhstan, China, Mongolia, the United States and Canada. In addition, it could be recreated at any location at which the gravitational acceleration had been accurately measured.

By the mid 19th century, increasingly accurate pendulum measurements by Edward Sabine and Thomas Young revealed that gravity, and thus the length of any pendulum standard, varied measurably with local geologic features such as mountains and dense subsurface rocks.So a pendulum length standard had to be defined at a single point on Earth and could only be measured there. This took much of the appeal from the concept, and efforts to adopt pendulum standards were abandoned.

The Metre

In the discussions leading up to the French adoption of the metric system in 1791, the leading candidate for the definition of the new unit of length, the metre, was the seconds pendulum at 45° North latitude. It was advocated by a group led by French politician Talleyrand and mathematician Antoine Nicolas Caritat de Condorcet. This was one of the three final options considered by the French Academy of Sciences committee. However, on March 19, 1791 the committee instead chose to base the metre on the length of the meridian through Paris. A pendulum definition was rejected because of its variability at different locations, and because it defined length by a unit of time. (However, since 1983 the metre has been officially defined in terms of the length of the second and the speed of light.) A possible additional reason is that the radical French Academy didn't want to base their new system on the second, a traditional and nondecimal unit from the *ancien regime.*

Although not defined by the pendulum, the final length chosen for the metre, 10^{-7} of the pole-to-equator meridian arc, was very close to the length of the seconds pendulum (0.9937 m), within 0.63%. Although no reason for this particular choice was given at the time, it was probably to facilitate the use of the seconds pendulum as a secondary standard, as was proposed in the official document. So the modern world's standard unit of length is certainly closely linked historically with the seconds pendulum.

5

Friction

Introduction

Friction is the force resisting the relative motion of solid surfaces, fluid layers, and material elements sliding against each other. There are several types of friction:

- Dry friction resists relative lateral motion of two solid surfaces in contact. Dry friction is subdivided into *static friction* ("stiction") between non-moving surfaces, and *kinetic friction* between moving surfaces.
- Fluid friction describes the friction between layers within a viscous fluid that are moving relative to each other.
- Lubricated friction is a case of fluid friction where a fluid separates two solid surfaces.
- Skin friction is a component of drag, the force resisting the motion of a solid body through a fluid.
- Internal friction is the force resisting motion between the elements making up a solid material while it undergoes deformation.

When surfaces in contact move relative to each other, the friction between the two surfaces converts kinetic energy into heat. This property can have dramatic consequences, as illustrated by the use of friction created by rubbing pieces of wood together to start a fire. Kinetic energy is converted to heat whenever motion with friction occurs, for example when a viscous fluid is stirred. Another important consequence of many types of friction can be wear, which may lead to performance degradation and/or damage to components. Friction is a component of the science of tribology.

Friction is not itself a fundamental force but arises from fundamental electromagnetic forces between the charged particles constituting the two contacting surfaces. The complexity of these interactions makes the calculation of friction from first principles impossible and necessitates the use of empirical methods for analysis and the development of theory.

History

The classic rules of sliding friction were discovered by Leonardo da Vinci (1452–1519), but remained unpublished in his notebooks. They were rediscovered by Guillaume Amontons (1699). Amontons presented the nature of friction in terms of surface irregularities and the force required to raise the weight pressing the surfaces together. This view was further elaborated by Belidor (representation of rough surfaces with spherical asperities, 1737) and Leonhard Euler (1750) who derived the angle of repose of a weight on an inclined plane and first distinguished between static and kinetic friction. A different explanation was provided by Desaguliers (1725), who demonstrated the strong cohesion forces between lead spheres of which a small cap is cut off and which were then brought into contact with each other.

The understanding of friction was further developed by Charles-Augustin de Coulomb (1785). Coulomb investigated the influence of four main factors on friction: the nature of the materials in contact and their surface coatings; the extent of the surface area; the normal pressure (or load); and the length of time that the surfaces remained in contact (time of repose). Coulomb further considered the influence of sliding velocity, temperature and humidity, in order to decide between the different explanations on the nature of friction that had been proposed. The distinction between static and dynamic friction is made in Coulomb's friction law, although this distinction was already drawn by Johann Andreas von Segner in 1758. The effect of the time of repose was explained by Musschenbroek (1762) by considering the surfaces of fibrous materials, with fibres meshing together, which takes a finite time in which the friction increases.

John Leslie (1766–1832) noted a weakness in the views of Amontons and Coulomb. If friction arises from a weight being drawn up the inclined plane of successive asperities, why isn't it balanced then through descending the opposite slope? Leslie was equally skeptical about the role of adhesion proposed by Desaguliers, which should on the whole have the same tendency to accelerate as to retard the motion. In his view friction should be seen as a time-dependent process of flattening, pressing down asperities, which creates new obstacles in what were cavities before.

Arthur Morrin (1833) developed the concept of sliding versus rolling friction. Osborne Reynolds (1866) derived the equation of viscous flow. This completed the classic empirical model of friction (static, kinetic, and fluid) commonly used today in engineering.

The focus of research during the last century has been to understand the physical mechanisms behind friction. F. Phillip Bowden and David Tabor (1950) showed that at a microscopic level, the actual area of contact between surfaces is a very small fraction of the apparent area. This actual area of contact, caused by "asperities" (roughness) increases with pressure, explaining the proportionality between normal force and frictional force. The development of the atomic force microscope (1986) has recently enabled scientists to study friction at the atomic scale.

Types of Friction

Dry Friction

Dry friction resists relative lateral motion of two solid surfaces in contact. The two regimes of dry friction are 'static friction' ("stiction") between non-moving surfaces, and *kinetic friction* (sometimes called sliding friction or dynamic friction) between moving surfaces.

Coulomb friction, named after Charles-Augustin de Coulomb, is an approximate model used to calculate the force of dry friction. It is governed by the equation:

$$F_f \leq \mu F_n$$

where

- F_f is the force of friction exerted by each surface on the other. It is parallel to the surface, in a direction opposite to the net applied force.
- μ is the coefficient of friction, which is an empirical property of the contacting materials,
- F_n is the normal force exerted by each surface on the other, directed perpendicular (normal) to the surface.

The Coulomb friction F_f may take any value from zero up to μF_n , and the direction of the frictional force against a surface is opposite to the motion that surface would experience in the absence of friction. Thus, in the static case, the frictional force is exactly what it must be in order to prevent motion between the surfaces; it balances the net force tending to cause such motion. In this case, rather than providing an estimate of the actual frictional force, the Coulomb approximation provides a threshold value for this force, above which motion would commence. This maximum force is known as traction.

The force of friction is always exerted in a direction that opposes movement (for kinetic friction) or potential movement (for static friction) between the two surfaces. For example, a curling stone sliding along the ice experiences a kinetic force slowing it down. For an example of potential movement, the drive wheels of an accelerating car experience a frictional force pointing forward; if they did not, the wheels would spin, and the rubber would slide backwards along the pavement. Note that it is not the direction of movement of the vehicle they oppose, it is the direction of (potential) sliding between tire and road.

Normal Force

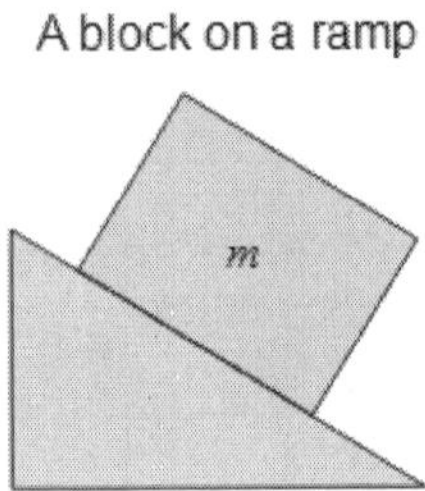

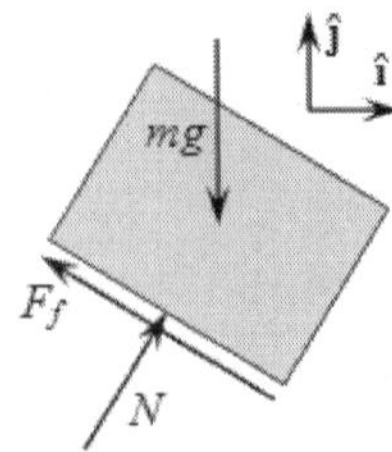

Figure: *Free-body diagram for a block on a ramp. Arrows are vectors indicating directions and magnitudes of forces.* N *is the normal force,* mg *is the force of gravity, and* F_f *is the force of friction.*

The normal force is defined as the net force compressing two parallel surfaces together; and its direction is perpendicular to the surfaces. In the simple case of a mass resting on a horizontal surface, the only component of the normal force is the force due to gravity, where $N = mg$. In this case, the magnitude of the friction force is the product of the mass of the object, the acceleration due to gravity, and the coefficient of friction. However, the coefficient of friction is not a function of mass or volume; it depends only on the material. For instance, a large aluminium block has the same coefficient of friction as a small aluminium block. However, the magnitude of the friction force itself depends on the normal force, and hence on the mass of the block.

If an object is on a level surface and the force tending to cause it to slide is horizontal, the normal force *N* between the object and the surface is just its weight, which is equal to its mass multiplied by the acceleration due to earth's gravity, *g*. If the object is on a tilted surface such as an inclined plane, the normal force is less, because less of the force of gravity is perpendicular to the face of the plane. Therefore, the normal force, and ultimately the frictional force, is determined using vector analysis, usually via a free body diagram. Depending on the situation, the calculation of the normal force may include forces other than gravity.

Coefficient of Friction

The coefficient of friction (COF), often symbolized by the Greek letter μ, is a dimensionless scalar value which describes the ratio of the force of friction between two bodies and the force pressing them together. The coefficient of friction depends on the materials used; for example, ice on steel has a low coefficient of friction, while rubber on pavement has a high coefficient of friction. Coefficients of friction range from near zero to greater than one.

For surfaces at rest relative to each other $\mu = \mu_s$, where μ_s is the *coefficient of static friction.* This is usually larger than its kinetic counterpart.

For surfaces in relative motion $\mu = \mu_k$, where μ_k is the *coefficient of kinetic friction.* The Coulomb friction is equal to F_f, and the frictional force on each surface is exerted in the direction opposite to its motion relative to the other surface.

Arthur Morin introduced the term and demonstrated the utility of the coefficient of friction. The coefficient of friction is an empirical measurement – it has to be measured experimentally, and cannot be found through calculations. Rougher surfaces tend to have higher effective values. Both static and kinetic coefficients of friction depend on the pair of surfaces in contact; for a given pair of surfaces, the coefficient of static friction is *usually* larger than that of kinetic friction; in some sets the two coefficients are equal, such as teflon-on-teflon.

Most dry materials in combination have friction coefficient values between 0.3 and 0.6. Values outside this range are rarer, but teflon, for example, can have a coefficient as low as 0.04. A value of zero would mean no friction at all, an elusive property – even magnetic levitation vehicles have drag. Rubber in contact with other surfaces can yield friction coefficients from 1 to 2. Occasionally it is maintained that μ is always < 1, but this is not true. While in most relevant applications $\mu < 1$, a value above 1 merely implies that the force required to slide an object along the surface is greater than the normal force of the surface on the object. For example, silicone rubber or acrylic rubber-coated surfaces have a coefficient of friction that can be substantially larger than 1.

While it is often stated that the COF is a "material property," it is better categorized as a "system property." Unlike true material properties (such as conductivity, dielectric constant, yield strength), the COF for any two materials depends on system variables like temperature, velocity, atmosphere and also what are now popularly described as aging and deaging times; as well as on geometric properties of the interface between the materials. For example, a copper pin sliding against a thick copper plate can have a COF that varies from 0.6 at low speeds (metal sliding against metal) to below 0.2 at high speeds when the copper surface begins to melt due to frictional heating. The latter speed, of course, does not determine the COF uniquely; if the pin diameter is increased so that the frictional heating is removed rapidly, the temperature drops, the pin remains solid and the COF rises to that of a 'low speed' test.

Negative Coefficient of Friction

As of 2012, a single study has demonstrated the potential for a negative coefficient of friction, meaning that a decrease in force leads to an increase in friction. This contradicts the everyday experience that an increase of normal force improves friction. This was reported in the journal *Nature* in October 2012 and involved the friction encountered by an atomic force microscope stylus when dragged across a graphene sheet in the presence of graphene-adsorbed oxygen.

Approximate Coefficients of Friction

Materials		*Static friction,*	
		Dry and clean	*Lubricated*
Aluminium	Steel	0.61	
Copper	Steel	0.53	
Brass	Steel	0.51	
Cast iron	Copper	1.05	
Cast iron	Zinc	0.85	
Concrete (wet)	Rubber	0.30	
Concrete (dry)	Rubber	1.0	
Concrete	Wood	0.62	
Copper	Glass	0.68	
Glass	Glass	0.94	
Metal	Wood	0.2–0.6	0.2 (wet)
Polyethene	Steel	0.2	0.2
Steel	Steel	0.80	0.16
Steel	PTFE	0.04	0.04
PTFE	PTFE	0.04	0.04
Wood	Wood	0.25–0.5	0.2 (wet)

An $AlMgB_{14}$-TiB_2 composite has an approximate coefficient of friction of 0.02 in water-glycol-based lubricants, and 0.04–0.05 when dry. Under certain conditions, some materials have even lower friction coefficients. An example is (highly ordered pyrolytic) graphite, which can have a friction coefficient below 0.01. This ultralow-friction regime is called superlubricity.

Static Friction

Static friction is friction between two or more solid objects that are not moving relative to each other. For example, static friction can prevent an object from sliding down a sloped surface. The coefficient of static friction, typically denoted as μ_s, is usually higher than the coefficient of kinetic friction.

The static friction force must be overcome by an applied force before an object can move. The maximum possible friction force between two surfaces before sliding begins is the product of the coefficient of static friction and the normal force: $F_{max} = \mu_s F_n$. When there is no sliding occurring, the friction force can have any value from zero up to F_{max} . Any force smaller than F_{max} attempting to slide one surface over the other is opposed by a frictional force of equal magnitude and opposite direction. Any force larger than F_{max} overcomes the force of static friction and causes sliding to occur. The instant sliding occurs, static friction is no longer applicable—the friction between the two surfaces is then called kinetic friction.

An example of static friction is the force that prevents a car wheel from slipping as it rolls on the ground. Even though the wheel is in motion, the patch of the tire in contact with the ground is stationary relative to the ground, so it is static rather than kinetic friction.

The maximum value of static friction, when motion is impending, is sometimes referred to as limiting friction, although this term is not used universally. It is also known as traction.

Kinetic Friction

Kinetic (or dynamic) friction occurs when two objects are moving relative to each other and rub together (like a sled on the ground). The coefficient of kinetic friction is typically denoted as μ_k, and is usually less than the coefficient of static friction for the same materials. However, Richard Feynman comments that "with dry metals it is very hard to show any difference."

New models are beginning to show how kinetic friction can be greater than static friction. Kinetic friction is now understood, in many cases, to be primarily caused by chemical bonding between the surfaces, rather than interlocking asperities; however, in many other cases roughness effects are dominant, for example in rubber to road friction. Surface roughness and contact area, however, do affect kinetic friction for micro- and nano-scale objects where surface area forces dominate inertial forces.

Angle of Friction

For the maximum angle of static friction between granular materials.

For certain applications it is more useful to define static friction in terms of the maximum angle before which one of the items will begin sliding. This is called the *angle of friction* or *friction angle*. It is defined as:

$$\tan\theta = \mu_s$$

where θ is the angle from vertical and μ_s is the static coefficient of friction between the objects. This formula can also be used to calculate μ_s from empirical measurements of the friction angle.

Friction at the Atomic Level

Determining the forces required to move atoms past each other is a challenge in designing nanomachines. In 2008 scientists for the first time were able to move a single atom across a surface, and measure the forces required. Using ultrahigh vacuum and nearly-zero temperature (5 K), a modified atomic force microscope was used to drag a cobalt atom, and a carbon monoxide molecule, across surfaces of copper and platinum.

Limitations of the Coulomb model

The Coulomb approximation mathematically follows from the assumptions that surfaces are in atomically close contact only over a small fraction of their overall area, that this contact area is proportional to the normal force (until saturation, which takes place when all area is in atomic contact), and that frictional force is proportional to the applied normal force, independently of the contact area (you can see the experiments on friction from Leonardo da Vinci). Such reasoning aside, however, the approximation is fundamentally an empirical construction. It is a rule of thumb describing the approximate outcome of an extremely complicated physical interaction. The strength of the approximation is its simplicity and versatility – though in general the relationship between normal force and frictional force is not exactly linear (and so the frictional force is not entirely independent of the contact area of the surfaces), the Coulomb approximation is an adequate representation of friction for the analysis of many physical systems.

When the surfaces are conjoined, Coulomb friction becomes a very poor approximation (for example, adhesive tape resists sliding even when there is no normal force, or a negative normal force). In this case, the frictional force may depend strongly on the area of contact. Some drag racing tires are adhesive in this way. However, despite the complexity of the fundamental physics behind friction, the relationships are accurate enough to be useful in many applications.

Numerical Simulation of the Coulomb Model

Despite being a simplified model of friction, the Coulomb model is useful in many numerical simulation applications such as multibody systems and granular material. Even its most simple expression encapsulates the fundamental effects of sticking and sliding which are required in many applied cases, although specific algorithms have to be designed in order to efficiently numerically integrate mechanical systems with Coulomb friction and bilateral and/or unilateral contact. Some quite nonlinear effects, such as the so-called Painlevé paradoxes, may be encountered with Coulomb friction.

Dry Friction and Instabilities

Dry friction can induce several types of instabilities in mechanical systems, which display a stable behaviour in the absence of friction. For instance, friction-related dynamical instabilities are thought to be responsible of brake squeal and of the 'song' of a glass harp, phenomena which involve stick and slip, modelled as a drop of friction coefficient with velocity.

A connection between dry friction and flutter instability in a simple mechanical system has been discovered.

Viscosity

The viscosity of a fluid is a measure of its resistance to gradual deformation by shear stress or tensile stress. For liquids, it corresponds to the informal notion of "thickness". For example, honey has a higher viscosity than water.

Viscosity is due to friction between neighbouring parcels of the fluid that are moving at different velocities. When fluid is forced through a tube, the fluid generally moves faster near the axis and very slowly near the walls, therefore some stress (such as a pressure difference between the two ends of the tube) is needed to overcome the friction between layers and keep the fluid moving. For the same velocity pattern, the stress required is proportional to the fluid's viscosity. A liquid's viscosity depends on the size and shape of its particles and the attractions between the particles.

A fluid that has no resistance to shear stress is known as an ideal fluid or inviscid fluid. Zero viscosity is observed only at very low temperatures, in superfluids. Otherwise all fluids have positive viscosity. If the viscosity is very high, for instance in pitch, the fluid will appear to be a solid in the short term. A liquid whose viscosity is less than that of water is sometimes known as a mobile liquid, while a substance with a viscosity substantially greater than water is called a viscous liquid.

Etymology

The word "viscosity" is derived from the Latin "viscum", meaning mistletoe. A viscous glue called birdlime was made from mistletoe berries and was used for lime-twigs to catch birds.

Definition

Shear viscosity:

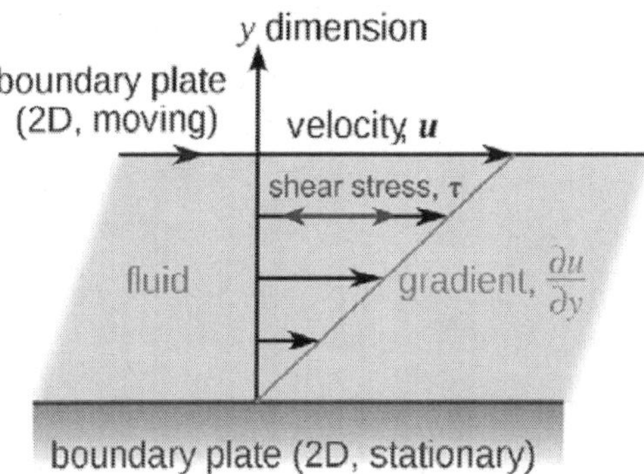

Figure: *Laminar shear of fluid between two plates. Friction between the fluid and the moving boundaries causes the fluid to shear. The force required for this action is a measure of the fluid's viscosity.*

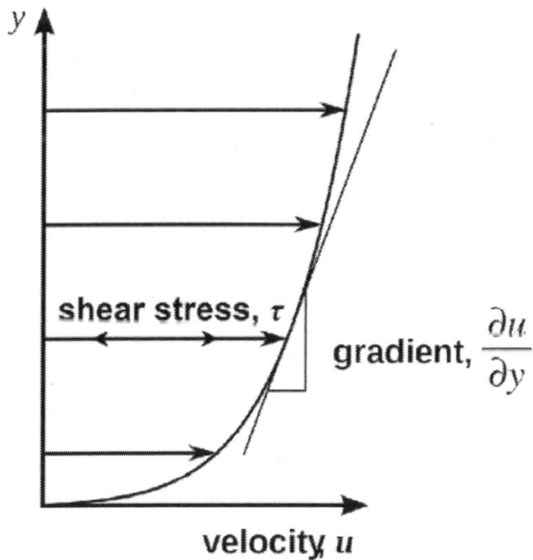

Figure: *In a general parallel flow (such as could occur in a straight pipe), the shear stress is proportional to the gradient of the velocity*

The shear viscosity of a fluid expresses its resistance to shearing flows, where adjacent layers move parallel to each other with different speeds. It can be defined through the idealized situation known as a Couette flow, where a layer of fluid is trapped between two horizontal plates, one fixed and one moving horizontally at constant speed u. (The plates are assumed to be very large, so that one need not consider what happens near their edges.)

If the speed of the top plate is small enough, the fluid particles will move parallel to it, and their speed will vary linearly from zero at the bottom to u at the top. Each layer of fluid will move faster than the one just below it, and friction between them will give rise to a force resisting their relative motion. In particular, the fluid will apply on the top plate a force in the direction opposite to its motion, and an equal but opposite one to the bottom plate. An external force is therefore required in order to keep the top plate moving at constant speed.

The magnitude F of this force is found to be proportional to the speed and the area A of each plate, and inversely proportional to their separation y. That is,

$$F = \mu A \frac{u}{y}$$

The proportionality factor μ in this formula is the viscosity (specifically, the dynamic viscosity) of the fluid.

The ratio u / y is called the *rate of shear deformation* or *shear velocity*, and is the derivative of the fluid speed in the direction perpendicular to the plates. Isaac Newton expressed the viscous forces by the differential equation

$$\tau = \mu \frac{\partial u}{\partial y}$$

where $\tau = F / A$ and $\partial u / \partial y$ is the local shear velocity. This formula assumes that the flow is moving along parallel lines and the y axis, perpendicular to the flow, points in the direction of maximum shear velocity. This equation can be used where the velocity does not vary linearly with y, such as in fluid flowing through a pipe.

Use of the Greek letter mu (μ) for the dynamic stress viscosity is common among mechanical and chemical engineers, as well as physicists. However, the Greek letter eta (η) is also used by chemists, physicists, and the IUPAC.

Kinematic Viscosity

The kinematic viscosity is the dynamic viscosity μ divided by the density of the fluid ρ. It is usually denoted by the Greek letter nu (ν). It is a convenient concept when analyzing the Reynolds number, that expresses the ratio of the inertial forces to the viscous forces:

$$Re = \frac{\rho u D}{\mu} = \frac{uD}{\nu}$$

Bulk Viscosity

When a compressible fluid is compressed or expanded evenly, without shear, it may still exhibit a form of internal friction that resists its flow. These forces are related to the rate of compression or expansion by a factor σ, called the volume viscosity, bulk viscosity or second viscosity. The bulk viscosity is important only when the fluid is being rapidly compressed or expanded, such as in sound and shock waves. Bulk viscosity explains the loss of energy in those waves, as described by Stokes' law of sound attenuation.

Viscosity Tensor

In general, the stresses within a flow can be attributed partly to the deformation of the material from some rest state (elastic stress), and partly to the rate of change of the deformation over time (viscous stress). In a fluid, by definition, the elastic stress includes only the hydrostatic pressure.

In very general terms, the fluid's viscosity is the relation between the strain rate and the viscous stress. In the Newtonian fluid model, the relationship is by definition a linear map, described by a viscosity tensor that, multiplied by the strain rate tensor (which is the gradient of the flow's velocity), gives the viscous stress tensor.

The viscosity tensor has nine independent degrees of freedom in general. For isotropic Newtonian fluids, these can be reduced to two independent parameters. The most usual decomposition yields the stress viscosity μ and the bulk viscosity σ.

Newtonian and Non-Newtonian Fluids

Newton's law of viscosity is a constitutive equation (like Hooke's law, Fick's law, Ohm's law): it is not a fundamental law of nature but an approximation that holds in some materials and fails in others.

A fluid that behaves according to Newton's law, with a viscosity μ that is independent of the stress, is said to be Newtonian. Gases, water and many common liquids can be considered Newtonian in ordinary conditions and contexts. There are many non-Newtonian fluids that significantly deviate from that law in some way or other. For example:

- Shear thickening liquids, whose viscosity increases with the rate of shear stress.
- Shear thinning liquids, whose viscosity decreases with the rate of shear stress.
- Thixotropic liquids, that become less viscous over time when shaken, agitated, or otherwise stressed.

- Rheopectic liquids, that become more viscous over time when shaken, agitated, or otherwise stressed.
- Bingham plastics that behave as a solid at low stresses but flows as a viscous fluid at high stresses.

Shear thinning liquids are very commonly, but misleadingly, described as thixotropic.

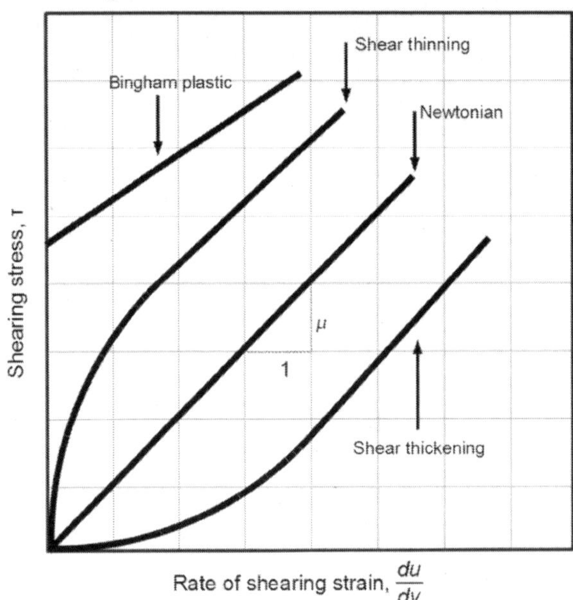

Figure: *Viscosity, the slope of each line, varies among materials*

Even for a Newtonian fluid, the viscosity usually depends on its composition and temperature. For gases and other compressible fluids, it depends on temperature and varies very slowly with pressure. The viscosity of some fluids may depend on other factors. A magnetorheological fluid, for example, becomes thicker when subjected to a magnetic field, possibly to the point of behaving like a solid.

Viscosity in Solids

The viscous forces that arise during fluid flow must not be confused with the elastic forces that arise in a solid in response shear, compression or extension stresses. While in the latter the stress is proportional to the *amount* of shear deformation, in a fluid it is proportional to the *rate* of deformation over time. (For this reason, Maxwell used the term fugitive elasticity for fluid viscosity.)

However, many liquids (including water) will briefly react like elastic solids when subjected to sudden stress. Conversely, many "solids" (even granite) will flow like liquids, albeit very slowly, even under arbitrarily small stress. Such materials are therefore best described as possessing both elasticity (reaction to deformation) and viscosity (reaction to rate of deformation); that is, being viscoelastic.

Indeed, some authors have claimed that amorphous solids, such as glass and many polymers, are actually liquids with a very high viscosity (e.g.~greater than 10^{12} Pa·s). However, other authors dispute this hypothesis, claiming instead that there is some threshold for the stress, below which most solids will not flow at all, and that alleged instances of glass flow in window panes of old buildings are due to the crude manufacturing process of older eras rather than to the viscosity of glass.

Viscoelastic solids may exhibit both stress viscosity and bulk viscosity. The extensional viscosity is a linear combination of the shear and bulk viscosities that describes the reaction of a solid elastic material to elongation. It is widely used for characterizing polymers.

In geology, earth materials that exhibit viscous deformation at least three times greater than their elastic deformation are sometimes called rheids.

Viscosity Measurement

Viscosity is measured with various types of viscometers and rheometers. A rheometer is used for those fluids that cannot be defined by a single value of viscosity and therefore require more parameters to be set and measured than is the case for a viscometer. Close temperature control of the fluid is essential to acquire accurate measurements, particularly in materials like lubricants, whose viscosity can double with a change of only 5 °C.

For some fluids, viscosity is a constant over a wide range of shear rates (Newtonian fluids). The fluids without a constant viscosity (non-Newtonian fluids) cannot be described by a single number. Non-Newtonian fluids exhibit a variety of different correlations between shear stress and shear rate.

One of the most common instruments for measuring kinematic viscosity is the glass capillary viscometer.

In paint industries, viscosity is commonly measured with a Zahn cup, in which the efflux time is determined and given to customers. The efflux time can also be converted to kinematic viscosities (centistokes, cSt) through the conversion equations. Also used in paint, a Stormer viscometer uses load-based rotation in order to determine viscosity. The viscosity is reported in Krebs units (KU), which are unique to Stormer viscometers.

A Ford viscosity cup measures the rate of flow of a liquid. This, under ideal conditions, is proportional to the kinematic viscosity. Vibrating viscometers can also be used to measure viscosity. These models such as the *Dynatrol* use vibration rather than rotation to measure viscosity.

Extensional viscosity can be measured with various rheometers that apply extensional stress.

Volume viscosity can be measured with an acoustic rheometer. Apparent viscosity is a calculation derived from tests performed on drilling fluid used in oil or gas well development. These calculations and tests help engineers develop and maintain the properties of the drilling fluid to the specifications required.

Units

Dynamic Viscosity: The SI physical unit of dynamic viscosity is the pascal-second (Pa·s), (equivalent to N·s/m^2, or kg/(m·s)). If a fluid with a viscosity of one Pa·s is placed between two plates, and one plate is pushed sideways with a shear stress of one pascal, it moves a distance equal to the thickness of the layer between the plates in one second. Water at 20 °C has a viscosity of 0.001002 Pa·s, and motor oil of about 0.250 Pa·s.

The cgs physical unit for dynamic viscosity is the *poise* (P), named after Jean Louis Marie Poiseuille. It is more commonly expressed, particularly in ASTM standards, as *centipoise* (cP). Water at 20 °C has a viscosity of 1.0020 cP.

1 P = 0.1 Pa·s,

1 cP = 1 mPa·s = 0.001 Pa·s.

Fluidity

The reciprocal of viscosity is *fluidity*, usually symbolized by $\varphi = 1 / \mu$ or $F = 1 / \mu$, depending on the convention used, measured in *reciprocal poise* (cm·s·g^{-1}), sometimes called the *rhe*. *Fluidity* is seldom used in engineering practice.

The concept of fluidity can be used to determine the viscosity of an ideal solution. For two components *a* and *b*, the fluidity when *a* and *b* are mixed is

$$F \approx \chi_a F_a + \chi_b F_b,$$

which is only slightly simpler than the equivalent equation in terms of viscosity:

$$\mu \approx \frac{1}{\chi_a / \mu_a + \chi_b / \mu_b},$$

where χ_a and χ_b is the mole fraction of component *a* and *b* respectively, and μ_a and μ_b are the components' pure viscosities.

Non-standard Units

The Reyn is a British unit of dynamic viscosity. Viscosity index is a measure for the change of kinematic viscosity with temperature. It is used to characterise lubricating oil in the automotive industry.

At one time the petroleum industry relied on measuring kinematic viscosity by means of the Saybolt viscometer, and expressing kinematic viscosity in units of *Saybolt Universal Seconds* (SUS). Other abbreviations such as SSU (*Saybolt Seconds Universal*) or SUV (*Saybolt Universal Viscosity*) are sometimes used. Kinematic viscosity in centistoke can be converted from SUS according to the arithmetic and the reference table provided in ASTM D 2161.

Molecular Origins

Figure: *Pitch has a viscosity approximately 230 billion (2.3×10^{11}) times that of water.*

The viscosity of a system is determined by how molecules constituting the system interact. There are no simple but correct expressions for the viscosity of a fluid. The simplest exact expressions are the Green–Kubo relations for the linear shear viscosity or the Transient Time Correlation Function expressions derived by Evans and Morriss in 1985. Although these expressions are each exact, in order to calculate the viscosity of a dense fluid using these relations currently requires the use of molecular dynamics computer simulations.

Gases

Viscosity in gases arises principally from the molecular diffusion that transports momentum between layers of flow. The kinetic theory of gases allows accurate prediction of the behaviour of gaseous viscosity.

Within the regime where the theory is applicable:

- Viscosity is independent of pressure and
- Viscosity increases as temperature increases.

James Clerk Maxwell published a famous paper in 1866 using the kinetic theory of gases to study gaseous viscosity. To understand why the viscosity is independent of pressure, consider two adjacent boundary layers (A and B) moving with respect to each other. The internal friction (the viscosity) of the gas is determined by the probability a particle of layer A enters layer B with a corresponding transfer of momentum. Maxwell's calculations show that the viscosity coefficient is proportional to the density, the mean free path, and the mean velocity of the atoms. On the other hand, the *mean free path* is inversely proportional to the density. So an increase in density due to an increase in pressure doesn't result in any change in viscosity.

Relation to Mean Free Path of Diffusing Particles

In relation to diffusion, the kinematic viscosity provides a better understanding of the behaviour of mass transport of a dilute species. Viscosity is related to shear stress and the rate of shear in a fluid, which illustrates its dependence on the mean free path, λ, of the diffusing particles.

From fluid mechanics, for a Newtonian fluid, the shear stress, τ, on a unit area moving parallel to itself, is found to be proportional to the rate of change of velocity with distance perpendicular to the unit area:

$$\tau = \mu \frac{du_x}{dy}$$

for a unit area parallel to the x-z plane, moving along the x axis. We will derive this formula and show how μ is related to λ.

Interpreting shear stress as the time rate of change of momentum, p, per unit area A (rate of momentum flux) of an arbitrary control surface gives

$$\tau = \frac{\dot{p}}{A} = \frac{\dot{m}\langle u_x \rangle}{A}.$$

where $\langle u_x \rangle$ is the average velocity, along the x axis, of fluid molecules hitting the unit area, with respect to the unit area.

Further manipulation will show

$$\dot{m} = \rho \bar{u} A$$

$\langle u_x \rangle = \frac{1}{2}\lambda \frac{\mathrm{d}u_x}{\mathrm{d}y}$, assuming that molecules hitting the unit area come from all distances between 0 and λ (equally distributed), and that their average velocities change linearly with distance (always true for small enough λ). From this follows:

$$\tau = \underbrace{\frac{1}{2}\rho\bar{u}\lambda}_{\mu} \cdot \frac{\mathrm{d}u_x}{\mathrm{d}y} \Rightarrow \nu = \frac{\mu}{\rho} = \tfrac{1}{2}\bar{u}\lambda,$$

where

$\dot{m}$ is the rate of fluid mass hitting the surface,

ρ is the density of the fluid,

$\hat{u}$ is the average molecular speed ($\bar{u} = \sqrt{\langle u^2 \rangle}$),

μ is the dynamic viscosity.

Effect of Temperature on the Viscosity of a Gas

Sutherland's formula can be used to derive the dynamic viscosity of an ideal gas as a function of the temperature:

$$\mu = \mu_0 \frac{T_0 + C}{T + C}\left(\frac{T}{T_0}\right)^{3/2}.$$

This in turn is equal to

$\lambda \frac{T^{3/2}}{T + C}$, where $\lambda = \frac{\mu_0 (T_0 + C)}{T_0^{3/2}}$ is a constant for the gas.

in Sutherland's formula:

- μ = dynamic viscosity in (Pa·s) at input temperature T,
- μ_0 = reference viscosity in (Pa·s) at reference temperature T_0,
- T = input temperature in kelvins,

- T_0 = reference temperature in kelvins,
- C = Sutherland's constant for the gaseous material in question.

Valid for temperatures between $0 < T < 555$ K with an error due to pressure less than 10% below 3.45 MPa.

According to Sutherland's formula, if the absolute temperature is less than C, the relative change in viscosity for a small change in temperature is greater than the relative change in the absolute temperature, but it is smaller when T is above C. The kinematic viscosity though always increases faster than the temperature (that is, d log(ν)/ d log(T) is greater than 1).

Sutherland's constant, reference values and λ values for some gases:

Gas	*C [K]*	T_0 *[K]*	μ_0 *[μPa s]*	λ *[μPa s* $K^{-1/2}$*]*
air	120	291.15	18.27	1.512041288
nitrogen	111	300.55	17.81	1.406732195
oxygen	127	292.25	20.18	1.693411300
carbon dioxide	240	293.15	14.8	1.572085931
carbon monoxide	118	288.15	17.2	1.428193225
hydrogen	72	293.85	8.76	0.636236562
ammonia	370	293.15	9.82	1.297443379
sulphur dioxide	416	293.65	12.54	1.768466086
helium	79.4	273	19	1.484381490

Viscosity of a Dilute Gas

The Chapman-Enskog equation may be used to estimate viscosity for a dilute gas. This equation is based on a semi-theoretical assumption by Chapman and Enskog. The equation requires three empirically determined parameters: the collision diameter (σ), the maximum energy of attraction divided by the Boltzmann constant ($^{\varepsilon}/\kappa$) and the collision integral ($\omega(T^*)$).

$$\mu_0 \times 10^6 = 2.6693 \frac{(MT)^{1/2}}{\sigma^2 \omega(T^*)},$$

with

- $T^* = \kappa T/\varepsilon$ — reduced temperature (dimensionless),
- μ_0 = viscosity for dilute gas (μPa.s),
- M = molecular mass (g/mol),
- T = temperature (K),

- σ = the collision diameter (Å),
- ε / κ = the maximum energy of attraction divided by the Boltzmann constant (K),
- ω_μ = the collision integral.

Liquids

Figure: *Video showing three liquids with different Viscosities*

In liquids, the additional forces between molecules become important. This leads to an additional contribution to the shear stress though the exact mechanics of this are still controversial. Thus, in liquids:

- Viscosity is independent of pressure (except at very high pressure); and
- Viscosity tends to fall as temperature increases (for example, water viscosity goes from 1.79 cP to 0.28 cP in the temperature range from 0 °C to 100 °C).

The dynamic viscosities of liquids are typically several orders of magnitude higher than dynamic viscosities of gases.

Viscosity of Blends of Liquids

The viscosity of the blend of two or more liquids can be estimated using the Refutas equation. The calculation is carried out in three steps.

The first step is to calculate the Viscosity Blending Number (VBN) (also called the Viscosity Blending Index) of each component of the blend:

(1) $$\text{VBN} = 14.534 \times \ln\left[\ln(\nu + 0.8)\right] + 10.975$$

where ν is the kinematic viscosity in centistokes (cSt). It is important that the kinematic viscosity of each component of the blend be obtained at the same temperature.

The next step is to calculate the VBN of the blend, using this equation:

(2) $$\text{VBN}_{\text{Blend}} = [x_A \times \text{VBN}_A] + [x_B \times \text{VBN}_B] + \cdots + [x_N \times \text{VBN}_N]$$

where x_X is the mass fraction of each component of the blend. Once the viscosity blending number of a blend has been calculated using equation (2), the final step is to determine the kinematic viscosity of the blend by solving equation (1) for ν:

(3) $$\nu = \exp\left(\exp\left(\frac{\text{VBN}_{\text{Blend}} - 10.975}{14.534}\right)\right) - 0.8,$$

where VBN_{Blend} is the viscosity blending number of the blend.

Viscosity of Selected Substances

Air:

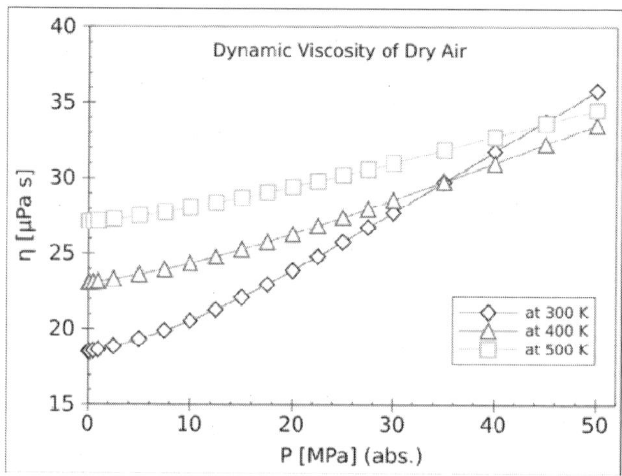

Figure: *Pressure dependence of the dynamic viscosity of dry air at the temperatures of 300, 400 and 500 K*

The viscosity of air depends mostly on the temperature. At 15 °C, the viscosity of air is 1.81×10^{-5} kg/(m·s), 18.1 μPa.s or 1.81×10^{-5} Pa.s. The kinematic viscosity at 15 °C is 1.48×10^{-5} m²/s or 14.8 cSt. At 25 °C, the viscosity is 18.6 μPa.s and the kinematic viscosity 15.7 cSt. One can get the viscosity of air as a function of temperature from the Gas Viscosity Calculator

Water

The dynamic viscosity of water is 8.90×10^{-4} Pa·s or 8.90×10^{-3} dyn·s/cm² or 0.890 cP at about 25 °C.

Water has a viscosity of 0.0091 poise at 25 °C, or 1 centipoise at 20 °C.

As a function of temperature T (K): (Pa·s) = $A \times 10^{B/(T*C)}$

where A=2.414 × 10^{-5} Pa·s ; B = 247.8 K ; and C = 140 K.

Viscosity of liquid water at different temperatures up to the normal boiling point is listed below.

Temperature [°C]	*Viscosity [mPa·s]*
10	1.308
20	1.002
30	0.7978
40	0.6531
50	0.5471
60	0.4658
70	0.4044
80	0.3550
90	0.3150
100	0.2822

Other Substances

Figure: *Example of the viscosity of milk and water. Liquids with higher viscosities make smaller splashes when poured at the same velocity.*

Figure: *Honey being drizzled.*

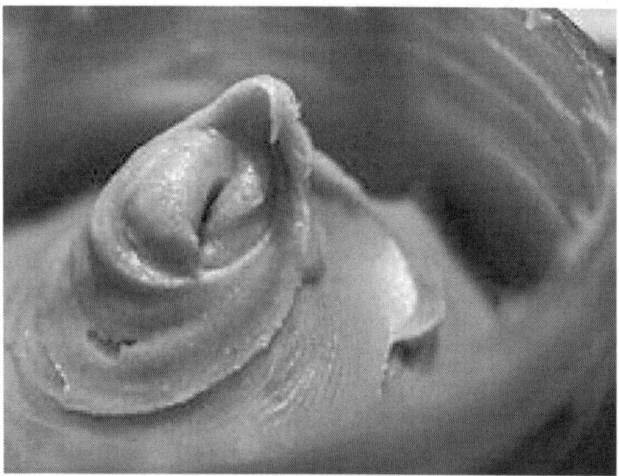

Figure: *Peanut butter is a semi-solid and can therefore hold peaks.*

Some dynamic viscosities of Newtonian fluids are listed below:

Viscosity of selected gases at 100 kPa, [μPa·s]		
Gas	***at 0 °C (273 K)***	***at 27 °C (300 K)***
air	17.4	18.6
hydrogen	8.4	9.0

Contd...

Gas	***at 0 °C (273 K)***	***at 27 °C (300 K)***
helium		20.0
argon		22.9
xenon	21.2	23.2
carbon dioxide		15.0
methane		11.2
ethane		9.5

Viscosity of fluids with variable compositions

Fluid	***Viscosity [Pa·s]***	***Viscosity [cP]***
blood (37 °C)	$(3–4)\times10^{-3}$	3–4
honey	2–10	2,000–10,000
molasses	5–10	5,000–10,000
molten glass	10–1,000	10,000–1,000,000
chocolate syrup	10–25	10,000–25,000
molten chocolate*	45–130	45,000–130,000
ketchup*	50–100	50,000–100,000
lard	≈ 100	≈ 100,000
peanut butter*	≈ 250	≈ 250,000
shortening*	≈ 250	≈ 250,000

Viscosity of liquids(at 25 °C unless otherwise specified)

Liquid :	***Viscosity [Pa·s]***	***Viscosity [cP=mPa·s]***
acetone	3.06×10^{-4}	0.306
benzene	6.04×10^{-4}	0.604
castor oil	0.985	985
corn syrup	1.3806	1380.6
ethanol	1.074×10^{-3}	1.074
ethylene glycol	1.61×10^{-2}	16.1
glycerol (at 20 °C)	1.2	1200
HFO-380	2.022	2022
mercury	1.526×10^{-3}	1.526
methanol	5.44×10^{-4}	0.544
motor oil SAE 10 (20 °C)	0.065	65
motor oil SAE 40 (20 °C)	0.319	319
nitrobenzene	1.863×10^{-3}	1.863

Contd...

Liquid :	***Viscosity [Pa·s]***	***Viscosity [cP=mPa·s]***
liquid nitrogen @ 77K	1.58×10^{-4}	0.158
propanol	1.945×10^{-3}	1.945
olive oil	.081	81
pitch	2.3×10^{8}	2.3×10^{11}
sulphuric acid	2.42×10^{-2}	24.2
water	8.94×10^{-4}	0.894

Viscosity of solids

Solid	***Viscosity [Pa·s]***	***Temperature [K]***
asthenosphere	7.0×10^{19}	900 °C
upper mantle	$(0.7\text{-}1.0)\times10^{21}$	1300-3000 °C
lower mantle	$(1.0\text{-}2.0)\times10^{21}$	3000-4000 °C

* *These materials are highly non-Newtonian.*

Note: *Higher viscosity means thicker substance*

Viscosity of Slurry

The term slurry describes mixtures of a liquid and solid particles that retain some fluidity. The viscosity of slurry can be described as relative to the viscosity of the liquid phase:

$$\mu_s = \mu_r \cdot \mu_l,$$

where μ_s and μ_l are respectively the dynamic viscosity of the slurry and liquid (Pa·s), and μ_r is the relative viscosity (dimensionless).

Depending on the size and concentration of the solid particles, several models exist that describe the relative viscosity as a function of volume fraction x of solid particles.

In the case of extremely low concentrations of fine particles, Einstein's equation may be used:

$$\mu_r = 1 + 2.5\cdot\phi$$

In the case of higher concentrations, a modified equation was proposed by Guth and Simha, which takes into account interaction between the solid particles:

$$\mu_r = 1 + 2.5\cdot\phi + 14.1\cdot\phi^2$$

Further modification of this equation was proposed by Thomas from the fitting of empirical data:

$$\mu_r = 1 + 2.5{\cdot}\phi + 10.05{\cdot}\phi^2 + A{\cdot}e^{B\cdot\phi},$$

where $A = 0.00273$ and $B = 16.6$.

In the case of very high concentrations, another empirical equation was proposed by Kitano *et al.*:

$$\mu_r = (1 - \frac{\phi}{A})^{-2},$$

where $A = 0.68$ for smooth spherical particles.

Viscosity of Amorphous Materials

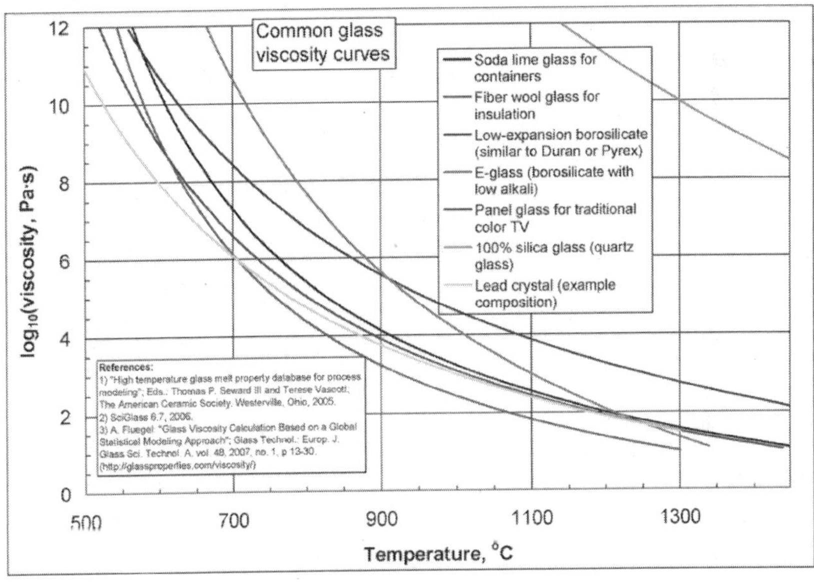

Figure: *Common glass viscosity curves.*

Viscous flow in amorphous materials (e.g. in glasses and melts) is a thermally activated process:

$$\mu = A{\cdot}e^{Q/RT},$$

where Q is activation energy, T is temperature, R is the molar gas constant and A is approximately a constant.

The viscous flow in amorphous materials is characterized by a deviation from the Arrhenius-type behaviour: Q changes from a high value Q_H at low temperatures (in the glassy state) to a low value Q_L at high temperatures (in the liquid state). Depending on this change, amorphous materials are classified as either

- strong when: $Q_H - Q_L < Q_L$ or
- fragile when: $Q_H - Q_L \geq Q_L$.

The fragility of amorphous materials is numerically characterized by the Doremus' fragility ratio:

$$R_D = \frac{Q_H}{Q_L}$$

and strong material have $R_D < 2$ whereas fragile materials have $R_D \geq 2$.

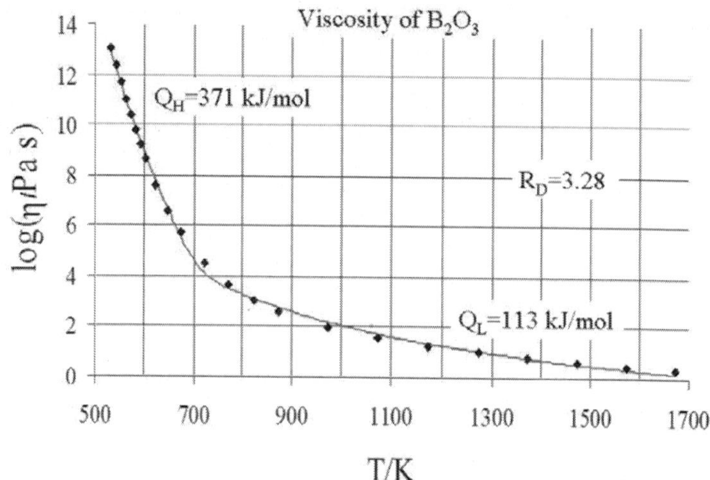

Figure: *Common log of viscosity vs temperature for* B_2O_3*, showing two regimes*

The viscosity of amorphous materials is quite exactly described by a two-exponential equation:

$$\mu = A_1 \cdot T \cdot \left[1 + A_2 \cdot e^{B/RT}\right] \cdot \left[1 + C \cdot e^{D/RT}\right],$$

with constants A_1, A_2, B, C and D related to thermodynamic parameters of joining bonds of an amorphous material.

Not very far from the glass transition temperature, T_g, this equation can be approximated by a Vogel-Fulcher-Tammann (VFT) equation.

If the temperature is significantly lower than the glass transition temperature, $T \ll T_g$, then the two-exponential equation simplifies to an Arrhenius type equation:

$$\mu = A_L T \cdot e^{Q_H/RT}$$

with:

$$Q_H = H_d + H_m,$$

where H_d is the enthalpy of formation of broken bonds (termed configuron s) and H_m is the enthalpy of their motion. When the temperature is less than the glass transition temperature, $T < T_g$, the activation energy of viscosity is high because the amorphous materials are in the glassy state and most of their joining bonds are intact.

If the temperature is highly above the glass transition temperature, $T \gg T_g$, the two-exponential equation also simplifies to an Arrhenius type equation:

$$\mu = A_H T \cdot e^{Q_L/RT},$$

with:

$$Q_L = H_m.$$

When the temperature is higher than the glass transition temperature, $T > T_g$, the activation energy of viscosity is low because amorphous materials are melted and have most of their joining bonds broken, which facilitates flow.

Eddy Viscosity

In the study of turbulence in fluids, a common practical strategy for calculation is to ignore the small-scale *vortices* (or *eddies*) in the motion and to calculate a large-scale motion with an *eddy viscosity* that characterizes the transport and dissipation of energy in the smaller-scale flow. Values of eddy viscosity used in modelling ocean circulation may be from $5\text{x}10^4$ to 10^6 Pa·s depending upon the resolution of the numerical grid.

Lubrication

Lubrication is the process, or technique employed to reduce wear of one or both surfaces in close proximity, and moving relative to each other, by interposing a substance called lubricant between the surfaces to carry or to help carry the load (pressure generated) between the opposing surfaces. The interposed lubricant film can be a solid, (e.g. graphite, MoS_2) a solid/liquid dispersion, a liquid, a liquid-liquid dispersion (a grease) or, exceptionally, a gas.

In the most common case the applied load is carried by pressure generated within the fluid due to the frictional viscous resistance to motion of the lubricating fluid between the surfaces.

Lubrication can also describe the phenomenon such reduction of wear occurs without human intervention (hydroplaning on a road).

The science of friction, lubrication and wear is called tribology.

Adequate lubrication allows smooth continuous operation of equipment, with only mild wear, and without excessive stresses or seizures at bearings. When lubrication breaks down, metal or other components can rub destructively over each other, causing destructive damage, heat, and failure.

The Regimes of Lubrication

As the load increases on the contacting surfaces three distinct situations can be observed with respect to the mode of lubrication, which are called regimes of lubrication:

- Fluid film lubrication is the lubrication regime in which through viscous forces the load is fully supported by the lubricant within the space or gap between the parts in motion relative to one another (the lubricated conjunction) and solid–solid contact is avoided.
  - Hydrostatic lubrication is when an external pressure is applied to the lubricant in the bearing, to maintain the fluid lubricant film where it would otherwise be squeezed out.
  - Hydrodynamic lubrication is where the motion of the contacting surfaces, and the exact design of the bearing is used to pump lubricant around the bearing to maintain the lubricating film. This design of bearing may wear when started, stopped or reversed, as the lubricant film breaks down.
- Elastohydrodynamic lubrication: Mostly for nonconforming surfaces or higher load conditions, the bodies suffer elastic strains at the contact. Such strain creates a load-bearing area, which provides an almost parallel gap for the fluid to flow through. Much as in hydrodynamic lubrication, the motion of the contacting bodies generates a flow induced pressure, which acts as the bearing force over the contact area. In such high pressure regimes, the viscosity of the fluid may rise considerably. At full elastohydrodynamic lubrication the generated lubricant film completely separates the surfaces. Contact between raised solid features, or *asperities*, can occur, leading to a mixed-lubrication or boundary lubrication regime.
- Boundary lubrication (also called boundary film lubrication): The bodies come into closer contact at their asperities; the heat developed by the local pressures causes a condition which is called stick-slip and some asperities break off. At the elevated temperature and pressure conditions chemically reactive

constituents of the lubricant react with the contact surface forming a highly resistant tenacious layer, or film on the moving solid surfaces (boundary film) which is capable of supporting the load and major wear or breakdown is avoided. Boundary lubrication is also defined as that regime in which the load is carried by the surface asperities rather than by the lubricant.

Besides supporting the load the lubricant may have to perform other functions as well, for instance it may cool the contact areas and remove wear products. While carrying out these functions the lubricant is constantly replaced from the contact areas either by the relative movement (hydrodynamics) or by externally induced forces.

Lubrication is required for correct operation of mechanical systems pistons, pumps, cams, bearings, turbines, cutting tools etc. where without lubrication the pressure between the surfaces in close proximity would generate enough heat for rapid surface damage which in a coarsened condition may literally weld the surfaces together, causing seizure.

In some applications, such as piston engines, the film between the piston and the cylinder wall also seals the combustion chamber, preventing combustion gases from escaping into the crankcase.

Parasitic Drag

Parasitic drag is drag caused by moving a solid object through a fluid medium (in the case of aerodynamics, more specifically, a gaseous medium). Parasitic drag is made up of many components, the most prominent being form drag. Skin friction and interference drag are also major components of parasitic drag.

In aviation, induced drag tends to be greater at lower speeds because a high angle of attack is required to maintain lift, creating more drag. However, as speed increases the induced drag becomes much less, but parasitic drag increases because the fluid is flowing faster around protruding objects increasing friction or drag. At even higher transonic and supersonic speeds, wave drag enters the picture. Each of these forms of drag changes in proportion to the others based on speed. The combined overall drag curve therefore shows a minimum at some airspeed - an aircraft flying at this speed will be at or close to its optimal efficiency. Pilots will use this speed to maximize the gliding range in case of an engine failure. However, to maximize the gliding endurance, the aircraft's speed would have to be at the point of minimum power, which occurs at lower speeds than minimum drag. At the point of minimum drag, $C_{D,o}$ (drag coefficient of aircraft when lift equals zero) is equal to $C_{D,i}$ (induced drag coefficient, or coefficient of drag created by lift). At the point of minimum power, $C_{D,o}$ is equal to one third times $C_{D,i}$. This can be proven by deriving the following equations:

$$F_{drag} = \frac{1}{2}\rho V^2 A_s C_D$$

and

$$C_D = C_{D,o} + C_{D,i}$$

where

$$C_{D,i} = KC_L^2$$

Form drag

Form drag or pressure drag arises because of the form of the object. The general size and shape of the body is the most important factor in form drag - bodies with a larger apparent cross-section will have a higher drag than thinner bodies. Sleek designs, or designs that are streamlined and change cross-sectional area gradually are also critical for achieving minimum form drag. Form drag follows the drag equation, meaning that it rises with the square of speed, and thus becomes more important for high speed aircraft.

Form drag depends on the longitudinal section of the body. A diligent choice of body profile is more than essential for low drag coefficient. Streamlines should be continuous and separation of the boundary layer with its attendant vortices should be avoided.

Profile Drag

Profile drag is usually defined as the sum of form drag and skin friction. However the term is sometimes used as a synonym for form drag.

Interference Drag

A characteristic that is dominant in bodies in transonic flow is the concept of interference drag. One can imagine two bodies of the aircraft (e.g. horizontal and vertical tail) that intersect at a particular point. Both bodies generate high supervelocities, possibly even supersonic. However, at the intersection there is less physical space for the flow to go and even higher supervelocities are generated resulting in much stronger local shock waves than would be expected if either one of the two bodies would be considered by itself. The stronger shock wave induces an increase in wave drag that is termed interference drag. Interference drag plays a role throughout the entire aircraft (e.g. nacelles, pylons, empennage) and its detrimental effect is always kept in mind by designers. Ideally, the pressure distributions on the intersecting bodies should complement each other's pressure distribution. If one body locally displays a negative pressure coefficient, the intersecting body should have positive pressure coefficient. In reality, however, this is not always possible. Particular geometric characteristics on aircraft often show how designers have dealt with the issue of interference drag. A prime example is the wing-body fairing which smooths the sharp angle between the wing and the fuselage. Another example is the junction between the horizontal and vertical tailplane in a T-tail. Often, an additional fairing (acorn) is positioned to reduce the added supervelocities. The position of the nacelle with respect to the wing is a third example of how interference-drag considerations dominate this geometric feature. For nacelles that are positioned beneath the wing, the lateral and longitudinal distance from the wing is dominated by interference-drag considerations. If there is little vertical space available between the wing and the nacelle (because of ground clearance) the nacelle is usually positioned much more in front of the wing.

Skin Friction

Skin friction arises from the friction of the fluid against the "skin" of the object that is moving through it. Skin friction arises from the interaction between the fluid and the skin of the body, and is directly related to the wetted surface, the area of the surface of the body that is in contact with the fluid. As with other components of parasitic drag, skin friction follows the drag equation and rises with the square of the velocity.

The skin friction coefficient, C_f, is defined by

$$C_f \equiv \frac{\tau_w}{\frac{1}{2}\rho U_\infty^2},$$

where τ_w is the local wall shear stress, ρ is the fluid density, and U_∞ is the free-stream velocity (usually taken outside of the boundary layer or at the inlet). It is related to the momentum thickness as

$$C_f = 2\frac{d\theta}{dx}.$$

For comparison, the turbulent empirical relation known as the 1/ 7 Power Law (derived by Theodore von Kármán) is:

$$C_f = \frac{0.0583}{Re^{0.2}},$$

where Re is the Reynolds number.

Skin friction is caused by viscous drag in the boundary layer around the object. The boundary layer at the front of the object is usually laminar and relatively thin, but becomes turbulent and thicker towards the rear. The position of the transition point depends on the shape of the object. There are two ways to decrease friction drag: the first is to shape the moving body so that laminar flow is possible, like an airfoil. The second method is to decrease the length and cross-section of the moving object as much as is practicable. To do so, a designer can consider the fineness ratio, which is the length of the aircraft divided by its diameter at the widest point (L/D).

Plasticity (physics)

In physics and materials science, plasticity describes the deformation of a material undergoing non-reversible changes of shape in response to applied forces. For example, a solid piece of metal being bent or pounded into a new shape displays plasticity as permanent changes occur within the material itself. In engineering, the transition from elastic behaviour to plastic behaviour is called yield.

Plastic deformation is observed in most materials including metals, soils, rocks, concrete, foams, bone and skin. However, the physical mechanisms that cause plastic deformation can vary widely. At the crystal scale, plasticity in metals is usually a consequence of dislocations. In most crystalline materials such defects are relatively rare. But there are also materials where defects are numerous and are part of the very crystal structure, in such cases plastic crystallinity can result. In brittle materials such as rock, concrete, and bone, plasticity is caused predominantly by slip at microcracks.

For many ductile metals, tensile loading applied to a sample will cause it to behave in an elastic manner. Each increment of load is accompanied by a proportional increment in extension, and when the load is removed, the piece returns exactly to its original size. However, once the load exceeds some threshold (the yield strength), the extension increases more rapidly than in the elastic region, and when the load is removed, some amount of the extension remains.

However, elastic deformation is an approximation and its quality depends on the considered time frame and loading speed. If the deformation behaviour includes elastic deformation as indicated in the adjacent graph it is also often referred to as elastic-plastic or elasto-plastic deformation.

Perfect plasticity is a property of materials to undergo irreversible deformation without any increase in stresses or loads. Plastic materials with hardening necessitate increasingly higher stresses to result in further plastic deformation. Generally plastic deformation is also dependent on the deformation speed, i.e. usually higher stresses have to be applied to increase the rate of deformation and such materials are said to deform visco-plastically.

Contributing Properties

The plasticity of a material is directly proportional to the ductility and malleability of the material.

Physical Mechanisms

***Figure:** Plasticity under a spherical Nanoindenter in (111) Copper. All particles in ideal lattice positions are omitted and the colour code refers to the von Mises stress field.*

Plasticity in Metals

Plasticity in a crystal of pure metal is primarily caused by two modes of deformation in the crystal lattice, slip and twinning. Slip is a shear deformation which moves the atoms through many interatomic distances relative to their initial positions. Twinning is the plastic deformation which takes place along two planes due to set of forces applied on a given metal piece. Most metals show more plasticity when hot than when cold. Lead shows sufficient plasticity at room temperature. But cast iron does not possess sufficient plasticity for any forging operation even when hot. This property is of importance in forming, shaping and extruding operations on metals. But most metals are rendered plastic by heating and hence shaped hot.

Slip Systems

Crystalline materials contain uniform planes of atoms organized with long-range order. Planes may slip past each other along their close-packed directions, as is shown on the slip systems wiki page. The result is a permanent change of shape within the crystal and plastic deformation. The presence of dislocations increases the likelihood of planes slipping.

Reversible Plasticity

On the nano scale the primary plastic deformation in simple fcc metals is reversible, as long as there is no material transport in form of cross-glide.

Shear Banding

The presence of other defects within a crystal may entangle dislocations or otherwise prevent them from gliding. When this happens, plasticity is localized to particular regions in the material. For crystals, these regions of localized plasticity are called shear bands.

Plasticity in Amorphous Materials

Crazing: In amorphous materials, the discussion of "dislocations" is inapplicable, since the entire material lacks long range order. These materials can still undergo plastic deformation. Since amorphous materials, like polymers, are not well-ordered, they contain a large amount of free volume, or wasted space. Pulling these materials in tension opens up these regions and can give materials a hazy appearance. This haziness is the result of *crazing*, where fibrils are formed within the material in regions of high hydrostatic stress. The material may go from an ordered appearance to a "crazy" pattern of strain and stretch marks.

Plasticity in Martensitic Materials

Some materials, especially those prone to Martensitic transformations, deform in ways that are not well described by the classic theories of plasticity and elasticity. One of the best-known examples of this is nitinol, which exhibits pseudoelasticity: deformations which are reversible in the context of mechanical design, but irreversible in terms of thermodynamics.

Plasticity in Cellular Materials

These materials plastically deform when the bending moment exceeds the fully plastic moment. This applies to open cell foams where the bending moment is exerted on the cell walls. The foams can be made of any material with a plastic yield point which includes rigid polymers and metals. This method of modelling the foam as beams is only valid if the ratio of the density of the foam to the density of the matter is less than 0.3. This is because beams yield axially instead of bending. In closed cell foams, the yield strength is increased if the material is under tension because of the membrane that spans the face of the cells.

Plasticity in Soils and Sand

Soils, particularly clays, display a significant amount of inelasticity under load. The causes of plasticity in soils can be quite complex and are strongly dependent on the microstructure, chemical composition, and water content. Plastic behaviour in soils is caused primarily by the rearrangement of clusters of adjacent grains.

Plasticity in Rocks and Concrete

Inelastic deformations of rocks and concrete are primarily caused by the formation of microcracks and sliding motions relative to these cracks. At high temperatures and pressures, plastic behaviour can also be affected by the motion of dislocations in individual grains in the microstructure.

Mathematical Descriptions of Plasticity

Deformation Theory: There are several mathematical descriptions of plasticity. One is deformation theory where the Cauchy stress tensor (of order d in d dimensions) is a function of the strain tensor. Although this description is accurate when a small part of matter is subjected to increasing loading (such as strain loading), this theory cannot account for irreversibility.

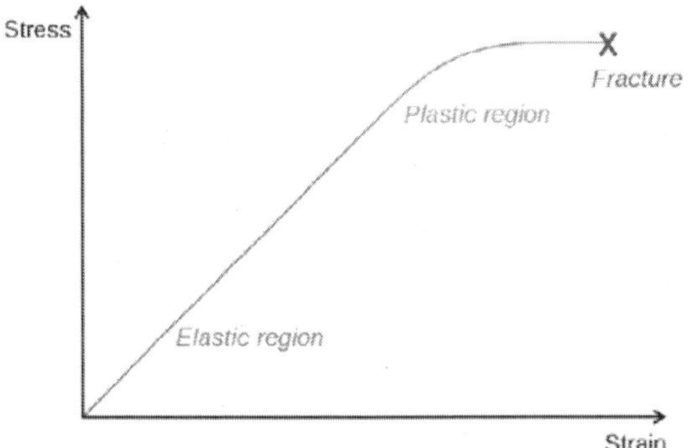

Figure: *An idealized uniaxial stress-strain curve showing elastic and plastic deformation regimes for the deformation theory of plasticity*

Ductile materials can sustain large plastic deformations without fracture.

However, even ductile metals will fracture when the strain becomes large enough - this is as a result of work hardening of the material, which causes it to become brittle.

Heat treatment such as annealing can restore the ductility of a worked piece, so that shaping can continue.

Flow Plasticity Theory

In 1934, Egon Orowan, Michael Polanyi and Geoffrey Ingram Taylor, roughly simultaneously, realized that the plastic deformation of ductile materials could be explained in terms of the theory of dislocations. The more correct mathematical theory of plasticity, flow plasticity theory, uses a set of non-linear, non-integrable equations to describe the set of changes on strain and stress with respect to a previous state and a small increase of deformation.

Yield Criteria

If the stress exceeds a critical value, as was mentioned above, the material will undergo plastic, or irreversible, deformation. This critical stress can be tensile or compressive. The Tresca and the von Mises criteria are commonly used to determine whether a material has yielded.

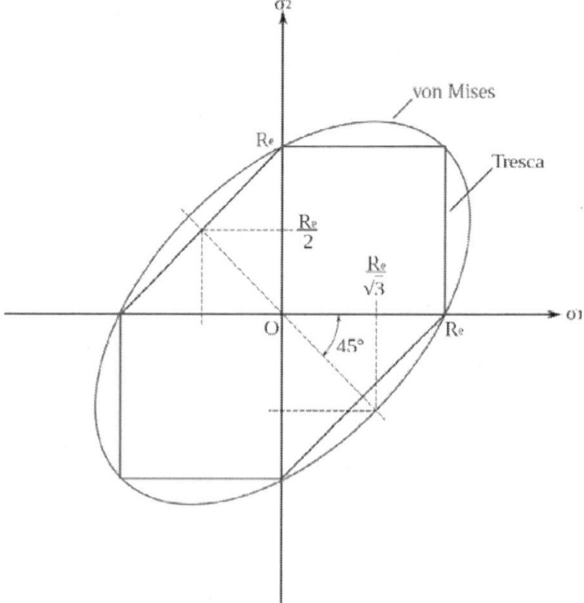

Figure: *Comparison of Tresca criterion to Von Mises criterion*

However, these criteria have proved inadequate for a large range of materials and several other yield criteria are in widespread use.

Tresca Criterion

This criterion is based on the notion that when a material fails, it does so in shear, which is a relatively good assumption when considering metals. Given the principal stress state, we can use Mohr's circle to solve for the maximum shear stresses our material will experience and conclude that the material will fail if:

$$\sigma_1 - \sigma_3 \geq \sigma_0$$

Where σ_1 is the maximum normal stress, σ_3 is the minimum normal stress, and σ_0 is the stress under which the material fails in uniaxial loading. A yield surface may be constructed, which provides a visual representation of this concept. Inside of the yield surface, deformation is elastic. On the surface, deformation is plastic. It is impossible for a material to have stress states outside its yield surface.

Huber-von Mises Criterion

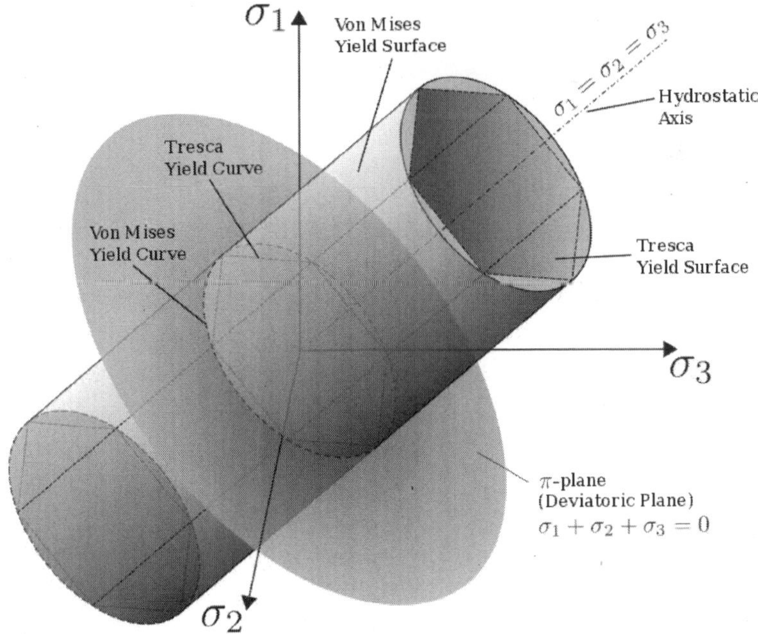

Figure: *The von Mises yield surfaces in principal stress coordinates circumscribes a cylinder around the hydrostatic axis. Also shown is Tresca's hexagonal yield surface.*

This criterion is based on the Tresca criterion but takes into account the assumption that hydrostatic stresses do not contribute to material failure. M.T. Huber was the first (1904, Lwów) who proposed the criterion of shear energy.

Von Mises solves for an effective stress under uniaxial loading, subtracting out hydrostatic stresses, and claims that all effective stresses greater than that which causes material failure in uniaxial loading will result in plastic deformation.

$$\sigma^2_{\text{effective}} = \tfrac{1}{2}\left[(\sigma_{11}-\sigma_{22})^2+(\sigma_{22}-\sigma_{33})^2+(\sigma_{11}-\sigma_{33})^2\right] + 6\left(\sigma_{12}^2+\sigma_{13}^2+\sigma_{23}^2\right)$$

Again, a visual representation of the yield surface may be constructed using the above equation, which takes the shape of an ellipse. Inside the surface, materials undergo elastic deformation. Reaching the surface means the material undergoes plastic deformations. It is physically impossible for a material to go beyond its yield surface.

Rolling Resistance

Rolling resistance, sometimes called rolling friction or rolling drag, is the force resisting the motion when a body (such as a ball, tire, or wheel) rolls on a surface. It is mainly caused by non-elastic effects; that is, not all the energy needed for deformation (or movement) of the wheel, roadbed, etc. is recovered when the pressure is removed. Two forms of this are hysteresis losses and permanent (plastic) deformation of the object or the surface (e.g. soil). Another cause of rolling resistance lies in the slippage between the wheel and the surface, which dissipates energy. Note that only the last of these effects involves friction, therefore the name "rolling friction" is to an extent a misnomer.

In analogy with sliding friction, rolling resistance is often expressed as a coefficient times the normal force. This coefficient of rolling resistance is generally much smaller than the coefficient of sliding friction.

Any coasting wheeled vehicle will gradually slow down due to rolling resistance including that of the bearings, but a train car with steel wheels running on steel rails will roll farther than a bus of the same mass with rubber tires running on tarmac. Factors that contribute to rolling resistance are the (amount of) deformation of the wheels, the deformation of the roadbed surface, and movement below the surface. Additional contributing factors include wheel diameter, speed, load on wheel, surface adhesion, sliding, and relative micro-sliding between the surfaces of contact. The losses due to hysteresis also depend strongly on the material properties of the wheel or tire and the surface. For example, a rubber tire will have higher rolling resistance on a paved road than a steel railroad wheel on a steel rail. Also, sand on the ground will give more rolling resistance than concrete.

Primary Cause

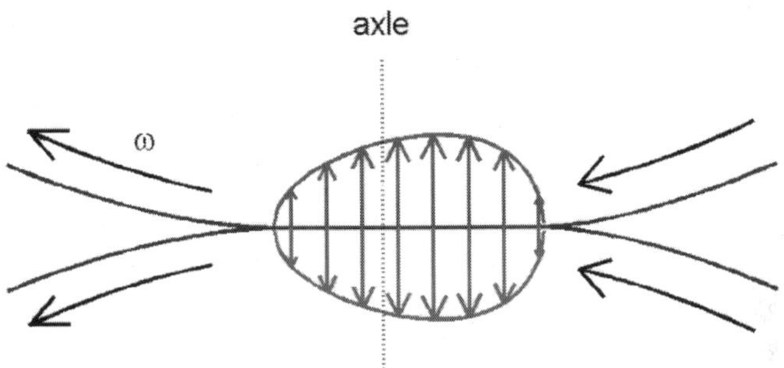

Figure: *Asymmetrical pressure distribution between rolling cylinders due to viscoelastic material behaviour (rolling to the right).*

The primary cause of pneumatic tire rolling resistance is hysteresis:

> *A characteristic of a deformable material such that the energy of deformation is greater than the energy of recovery. The rubber compound in a tire exhibits hysteresis. As the tire rotates under the weight of the vehicle, it experiences repeated cycles of deformation and recovery, and it dissipates the hysteresis energy loss as heat. Hysteresis is the main cause of energy loss associated with rolling resistance and is attributed to the viscoelastic characteristics of the rubber.— National Academy of Sciences*

This main principle is illustrated in the figure of the rolling cylinders. If two equal cylinders are pressed together then the contact surface is flat. In the absence of surface friction, contact stresses are normal (i.e. perpendicular) to the contact surface. Consider a particle that enters the contact area at the right side, travels through the contact patch and leaves at the left side. Initially its vertical deformation is increasing, which is resisted by the hysteresis effect. Therefore an additional pressure is generated to avoid interpenetration of the two surfaces. Later its vertical deformation is decreasing. This is again resisted by the hysteresis effect. In this case this decreases the pressure that is needed to keep the two bodies separate.

The resulting pressure distribution is asymmetrical and is shifted to the right. The line of action of the (aggregate) vertical force no longer passes through the centres of the cylinders. This means that a moment occurs that tends to retard the rolling motion.

Materials that have a large hysteresis effect, such as rubber, which bounce back slowly, exhibit more rolling resistance than materials with a small hysteresis effect that bounce back more quickly and more completely, such as steel or silica. Low rolling resistance tires typically incorporate silica in place of carbon black in their tread compounds to reduce low-frequency hysteresis without compromising traction. Note that railroads also have hysteresis in the roadbed structure.

"Rolling Resistance" has Different Definitions

In the broad sense, specific "rolling resistance" (for vehicles) is the force per unit vehicle weight required to move the vehicle on level ground at a constant slow speed where aerodynamic drag (air resistance) is insignificant and also where there are no traction (motor) forces or brakes applied. In other words the vehicle would be coasting if it were not for the force to maintain constant speed. An example of such usage for railroads is [3]. This broad sense includes wheel bearing resistance, the energy dissipated by vibration and oscillation of both the roadbed and the vehicle, and sliding of the wheel on the roadbed surface (pavement or a rail).

But there is an even broader sense which would include energy wasted by wheel slippage due to the torque applied from the engine. This includes the increased power required due to the increased velocity of the wheels where the tangential velocity of the driving wheel(s) becomes greater than the vehicle speed due to slippage. Since power is equal to force times velocity and the wheel velocity has increased, the power required has increased accordingly.

The pure "rolling resistance" for a train is that which happens due to deformation and possible minor sliding at the wheel-road contact. For a rubber tire, an analogous energy loss happens over the entire tire, but it is still called "rolling resistance". In the broad sense, "rolling resistance" includes wheel bearing resistance, energy loss by shaking both the roadbed (and the earth underneath) and the vehicle itself, and by sliding of the wheel, road/rail contact. Railroad textbooks seem to cover all these resistance forces but do not call their sum "rolling resistance" (broad sense) as is done in this article. They just sum up all the resistance forces (including aerodynamic drag) and call the sum basic train resistance (or the like).

Since railroad rolling resistance in the broad sense may be a few times larger than just the pure rolling resistance reported values may be in serious conflict since they may be based on different definitions of "rolling resistance". The train's engines must of course, provide the energy to overcome this broad-sense rolling resistance. For highway motor vehicles, there is obviously some energy dissipating in the shaking the roadway and earth beneath, shaking of the vehicle itself, and sliding of the tires. But other than the additional power required due to torque and wheel bearing friction, non-pure rolling resistance doesn't seem to have been investigated, possibly because the "pure" rolling resistance of a rubber tire is several times higher than the neglected resistances.

Rolling Resistance Coefficient

The "rolling resistance coefficient", is defined by the following equation:

$$F = C_{rr} N$$

where

F is the rolling resistance force,

C_{rr} is the dimensionless rolling resistance coefficient or coefficient of rolling friction (CRF), and

N is the normal force, the force perpendicular to the surface on which the wheel is rolling.

C_{rr} is the force needed to push (or tow) a wheeled vehicle forward (at constant speed on the level with no air resistance) per unit force of weight. It's assumed that all wheels are the same and bear identical weight. Thus: $C_{rr} = 0.01$ means that it would only take 0.01 pound to tow a vehicle weighing one pound. For a 1000 pound vehicle it would take 1000 times more tow force or 10 pounds. One could say that is C_{rr} in lb(tow-force)/lb(vehicle weight. Since this lb/lb is force divided by force, C_{rr} is dimensionless. Multiply it by 100 and you get the percent (%)of the weight of the vehicle required to maintain slow steady speed. C_{rr} is often multiplied by 1000 to get the parts per thousand which is the same as kilograms (kg force) per metric ton (tonne = 1000 kg) which is the same as pounds of resistance per 1000 pounds of load or Newtons/kilo-Newton, etc. For the US railroads, lb/ton has been traditionally used which is just $2000C_{rr}$. Thus they are all just measures of resistance per unit vehicle weight. While they are all "specific resistances" sometimes they are just called "resistance" although they are really a coefficient (ratio) or a multiple thereof. If using pounds or kilograms as force units, mass is equal to weight (in earth's gravity a kilogram a mass weighs a kilogram and exerts a kilogram of force) so one could claim that C_{rr} is also the force per unit mass in such units. The SI system would use

N/tonne (N/T) which is $1000gC_{rr}$ and is force per unit mass, where g is the acceleration of gravity in SI units (meters per second square).

The above shows resistance proportional to C_{rr} but does not explicitly show any variation with speed, loads, torque, surface roughness, diameter, tire inflation/wear, etc. because C_{rr} itself varies with those factors. It might seem from the above definition of C_{rr} that the rolling resistance is directly proportional to vehicle weight but it is not.

Measurement

There are at least two popular models for calculating rolling resistance.

1. "Rolling resistance coefficient (RRC). The value of the rolling resistance force divided by the wheel load. The Society of Automotive Engineers (SAE) has developed test practices to measure the RRC of tires. These tests (SAE J1269 and SAE J2452) are usually performed on new tires. When measured by using these standard test practices, most new passenger tires have reported RRCs ranging from 0.007 to 0.014." In the case of bicycle tires, values of 0.0025 to 0.005 are achieved. These coefficients are measured on rollers, with power meters on road surfaces, or with coast-down tests. In the latter two cases, the effect of air resistance must be subtracted or the tests performed at very low speeds.
2. The coefficient of rolling resistance b, which has the dimension of length, is approximately (due to the small-angle approximation of $cos(\theta) = 1$) equal to the value of the rolling resistance force times the radius of the wheel divided by the wheel load.
3. ISO 18164:2005 is used to test rolling resistance in Europe.

The results of these tests can be hard for the general public to obtain as manufacturers prefer to publicize "comfort" and "performance".

Physical Formulas

The coefficient of rolling friction for a slow rigid wheel on a perfectly elastic surface, not adjusted for velocity, can be calculated by

$$C_{rr} = \sqrt{z / d}$$

where

z is the sinkage depth

d is the diameter of the rigid wheel

Empirical formula for Crr for cast iron mine car wheels on steel rails.

$$C_{rr} = 0.0048(18/D)^{\frac{1}{2}}(100/W)^{\frac{1}{4}}$$

where

$D$is the wheel diameter in in.

W is the load on the wheel in lbs.

As an alternative to using C_{rr} one can use b which is a different rolling resistance coefficient or coefficient of rolling friction with dimension of length, It's defined by the following formula:

$$F = \frac{Nb}{r}$$

where

F is the rolling resistance force,

r is the wheel radius,

$b$is the rolling resistance coefficient or coefficient of rolling friction with dimension of length, and

$N$is the normal force (equal to W, not R).

The above equation, where resistance is inversely proportional to radius r. seems to be based on the discredited "Coulomb's law". Equating this equation with the force per the #Rolling resistance coefficient, and solving for b, gives b = C_{rr} r. Therefore, if a source gives rolling resistance coefficient (C_{rr}) as a dimensionless coefficient, it can be converted to b, having units of length, by multiplying C_{rr} by wheel radius r.

Rolling Resistance Coefficient Examples

Table of rolling resistance coefficient examples:

C_{rr}	*b*	*Description*
0.0003 *to* 0.0004		"Pure rolling resistance" Railroad steel wheel on steel rail
0.0010 *to* 0.0024	0.5 mm	Railroad steel wheel on steel rail. Passenger rail car about 0.0020
0.001 *to* 0.0015	0.1 mm	Hardened steel ball bearings on steel
0.0019 *to* 0.0065		Mine car cast iron wheels on steel rail
0.0022 *to* 0.005		Production bicycle tires at 120 psi (8.3 bar) and 50 km/h (31 mph), measured on rollers

Contd...

C_{rr}	*b*	*Description*
0.0025		Special Michelin solar car/eco-marathon tires
0.005		Dirty tram rails (standard) with straights and curves
0.0045 *to* 0.008[24]		Large truck (Semi) tires
0.0055		Typical BMX bicycle tires used for solar cars
0.0062 *to* 0.015		Car tire measurements
0.010 *to* 0.015		Ordinary car tires on concrete
0.0385 *to* 0.073		Stage coach (19th century) on dirt road. Soft snow on road for worst case.
0.3		Ordinary car tires on sand

For example, in earth gravity, a car of 1000 kg on asphalt will need a force of around 100 newtons for rolling (1000 kg × 9.81 m/s^2 × 0.01 = 98.1 N).

Depends on Diameter

Stagecoaches and Railroads (diameter): According to Dupuit (1837), rolling resistance (of wheeled carriages with wooden wheels with iron tires) is approximately inversely proportional to the square root of wheel diameter. This rule has been experimentally verified for cast iron wheels (8" - 24" diameter) on steel rail and for 19th century carriage wheels. But there are other tests of carriage wheels that do not agree. Theory of a cylinder rolling on an elastic roadway also gives this same rule These contradict earlier (1785) tests by Coulomb of rolling wooden cylinders where Coulomb reported that rolling resistance was inversely proportional to the diameter of the wheel (known as "Coulomb's law"). This disputed (or wrongly applied) -"Coulomb's law" is still found in handbooks, however.

Pneumatic Tires (diameter)

For pneumatic tires on hard pavement, it is reported that the effect of diameter on rolling resistance is negligible (within a practical range of diameters). This is another example of the inapplicability of "Coulomb's law".

Depends on Applied Torque

The driving torque T to overcome rolling resistance R_r and maintain steady speed on level ground (with no air resistance) can be calculated by:

$$T = \frac{V_s}{\Omega} R_r$$

where

V_s is the linear speed of the body (at the axle), and

Ω its rotational speed.

It is noteworthy that V_s / Ω is usually not equal to the radius of the rolling body.

All Wheels (torque)

"Applied torque" may either be driving torque applied by a motor (often through a transmission) or a braking torque applied by brakes (including regenerative braking). Such torques results in energy dissipation (above that due to the basic rolling resistance of a freely rolling, non-driven, non-braked wheel). This additional loss is in part due to the fact that there is some slipping of the wheel, and for pneumatic tires, there is more flexing of the sidewalls due to the torque. Slip is defined such that a 2% slip means that the circumferential speed of the driving wheel exceeds the speed of the vehicle by 2%.

A small percentage slip can result in a much larger percentage increase in rolling resistance. For example, for pneumatic tires, a 5% slip can translate into a 200% increase in rolling resistance. This is partly because the tractive force applied during this slip is many times greater than the rolling resistance force and thus much more power per unit velocity is being applied (recall power = force x velocity so that power per unit of velocity is just force).

So just a small percentage increase in circumferential velocity due to slip can translate into a loss of traction power which may even exceed the power loss due to basic (ordinary) rolling resistance. For railroads, this effect may be even more pronounced due to the low rolling resistance of steel wheels.

Railroad Steel Wheels (torque)

In order to apply any traction to the wheels some slippage of the wheel is required. For Russian trains climbing up a grade, this slip is normally 1.5% to 2.5%.

Slip (also known as creep)is normally roughly directly proportional to tractive effort. An exception is if the tractive effort is so high that the wheel is close to substantial slipping (more than just a few percent as discussed above), then slip rapidly increases with tractive effort and is no longer linear. With a little higher applied tractive effort the wheel spins out of control and the adhesion drops resulting in the wheel spinning even faster. This is the type of slipping that is observable by eye—the slip of say 2% for traction is only observed by instruments. Such rapid slip may result in excessive wear or damage.

Pneumatic Tires (torque)

Rolling resistance greatly increases with applied torque. At high torques, which apply a tangential force to the road of about half the weight of the vehicle, the rolling resistance may triple (a 200% increase). This is in part due to a slip of about 5%. The rolling resistance increase with applied torque is not linear, but increases at a faster rate as the torque becomes higher.

Depends on Wheel Load

Railroad Steel Wheels (load): The #Rolling resistance coefficient, Crr, significantly decreases as the weight of the rail car per wheel increases. For example, an empty Russian freight car had about twice the Crr as loaded car (Crr=0.002 vs. Crr=0.001). This same "economy of scale" shows up in testing of mine rail cars. The theoretical Crr for a rigid wheel rolling on an elastic roadbed shows Crr inversely proportional to the square root of the load.

If Crr is itself dependent on wheel load per an inverse square-root rule, then for an increase in load of 2% only a 1% increase in rolling resistance occurs.

Pneumatic Tires (load)

For pneumatic tires, the direction of change in Crr (#Rolling resistance coefficient) depends on whether or not tire inflation is increased with increasing load. It's reported that if inflation pressure is increased with load according to an (undefined) "schedule", then a 20% increase in load decreases Crr by 3%. But if the inflation pressure is not changed, then a 20% increase in load results in a 4% increase in Crr. Of course this will increase the rolling resistance by 20% due to the increase in load plus 1.2 x 4% due to the increase in Crr resulting in a 24.8% increase in rolling resistance.

Depends on Curvature of Roadway

General: When a vehicle (motor vehicle or railroad train) goes around a curve, rolling resistance usually increases. If the curve is not banked so as to exactly counter the centrifugal force with an equal and opposing centripetal force due to the banking, then there will be a net unbalanced sideways force on the vehicle.

This will result in increased rolling resistance. Banking is also known as "superelevation" or "cant" (not to be confused with rail cant of a rail). For railroads, this is called curve resistance but for roads it has (at least once) been called rolling resistance due to cornering.

Sound Effects

Rolling friction generates sound (vibrational) energy, as mechanical energy is converted to this form of energy due to the friction. One of the most common examples of rolling friction is the movement of motor vehicle tires on a roadway, a process which generates sound as a by-product.

The sound generated by automobile and truck tires as they roll (especially noticeable at highway speeds) is mostly due to the percussion of the tire treads, and compression (and subsequent decompression) of air temporarily captured within the treads.

Factors that Contribute in Tires

Several factors affect the magnitude of rolling resistance a tire generates:

- As mentioned in the introduction: wheel radius, forward speed, surface adhesion, and relative micro-sliding.
- Material - different fillers and polymers in tire composition can improve traction while reducing hysteresis. The replacement of some carbon black with higher-priced silica–silane is one common way of reducing rolling resistance. The use of exotic materials including nano-clay has been shown to reduce rolling resistance in high performance rubber tires. Solvents may also be used to swell solid tires, decreasing the rolling resistance.
- Dimensions - rolling resistance in tires is related to the flex of sidewalls and the contact area of the tire For example, at the same pressure, wider bicycle tires flex less in sidewalls as they roll and thus have lower rolling resistance (although higher air resistance).
- Extent of inflation - Lower pressure in tires results in more flexing of sidewalls and higher rolling resistance. This energy conversion in the sidewalls increases resistance and can also lead to overheating and may have played a part in the infamous Ford Explorer rollover accidents.
- Over inflating tires (such a bicycle tires) may not lower the overall rolling resistance as the tire may skip and hop over the road surface. Traction is sacrificed, and overall rolling friction

may not be reduced as the wheel rotational speed changes and slippage increases.

- Sidewall deflection is not a direct measurement of rolling friction. A high quality tire with a high quality (and supple) casing will allow for more flex per energy loss than a cheap tire with a stiff sidewall. Again, on a bicycle, a quality tire with a supple casing will still roll easier than a cheap tire with a stiff casing. Similarly, as noted by Goodyear truck tires, a tire with a "fuel saving" casing will benefit the fuel economy through many tread lives (i.e. retreading), while a tire with a "fuel saving" tread design will only benefit until the tread wears down.
- In tires, tread thickness and shape has much to do with rolling resistance. The thicker and more contoured the tread, the higher the rolling resistance Thus, the "fastest" bicycle tires have very little tread and heavy duty trucks get the best fuel economy as the tire tread wears out.
- Diameter effects seem to be negligible, provided the pavement is hard and the range of diameters is limited.
- Virtually all world speed records have been set on relatively narrow wheels, probably because of their aerodynamic advantage at high speed, which is much less important at normal speeds.
- Temperature: with both solid and pneumatic tires, rolling resistance has been found to decrease as temperature increases (within a range of temperatures: i.e. there is an upper limit to this effect) For a rise in temperature from 30 deg. C to 70 deg. C the rolling resistance decreased by 20-25% It's claimed that racers heat their tire before racing.

Railroads: Components of Rolling Resistance

One may define rolling resistance in the broad sense as the sum of components): 1. Wheel bearing resistance 2. Pure rolling resistance 3. Sliding of the wheel on the rail 4. Loss of energy to the roadbed (and earth) 5. Loss of energy to oscillation of railway rolling stock.

The wheel bearing resistance may be reported as a specific resistance at the wheel rim, for example as a Crr. Railroads normally use roller bearings which are either cylindrical (Russia) or tapered (United States). The specific rolling resistance (in of Russian bearings varies with both wheel loading and speed. It is lowest with high axle loads and intermediate speeds of 60–80 km/h with a Crr of 0.00013 (axle load of 21 tonnes). For empty freight cars with axle loads of 5.5 tonnes, Crr goes up to 0.00020 at 60 km/h but at a low speed of 20 km/h it increases to 0.00024 and at a high speed (for freight trains) of 120 km/h it is 0.00028. The Crr obtained above is added to the Crr of the other components to obtain the total Crr for the wheels.

Comparing Rolling Resistance of Highway Vehicles and Trains

While the specific rolling resistance of a train is far less than an automobile or truck in terms of resistance force per ton, this does not necessarily mean that the resistance force per passenger or per net ton of freight is less. It all depends on the vehicle weight per passenger or per net ton transported. Thus one needs to know the rolling resistance per passenger (or per net ton) to make such comparisons.

For 1975, Amtrak passenger trains weighed a little over 7 tonnes per passenger while automobiles weighed only a little over one ton per passenger. To find the rolling resistance per person one multiples the pounds (force) per ton (2000 times the rolling resistance coefficient) by the tons per passenger. This means that even if the rolling coefficient is several times greater for the auto than for the train, then after multiplication to get pounds/passenger, there is not a lot of difference between the two values (of lb/passenger). Thus there may not be a large difference in the rolling resistance energy used to transport a person by rail as compared to auto.

Triboelectric Effect

The triboelectric effect (also known as *triboelectric charging*) is a type of contact electrification in which certain materials become electrically charged after they come into contact with another different material through friction. Rubbing glass with fur, or a comb through the hair, can build up triboelectricity. Most everyday static electricity is triboelectric. The polarity and strength of the charges produced differ according to the materials, surface roughness, temperature, strain, and other properties.

Thus, it is not very predictable, and only broad generalizations can be made. Amber, for example, can acquire an electric charge by contact and separation (or friction) with a material like wool. This property, first recorded by Thales of Miletus, suggested the word "electricity" (from William Gilbert's initial coinage, "electra"), from the Greek word for amber, *çlektron*. Other examples of materials that can acquire a significant charge when rubbed together include glass rubbed with silk, and hard rubber rubbed with fur.

Triboelectric Series

John Carl Wilcke published the first triboelectric series in a 1757 paper on static charges. Materials are often listed in order of the polarity of charge separation when they are touched with another object. A material towards the bottom of the series, when touched to a material near the top of the series, will attain a more negative charge. The further away two materials are from each other on the series, the greater the charge transferred. Materials near to each other on the series may not exchange any charge, or may exchange the opposite of what is implied by the list. This depends more on the presence of rubbing, the presence of contaminants or oxides, or upon properties other than on the type of material. Lists vary somewhat as to the exact order of some materials, since the charge also varies for nearby materials. From actual tests, there is little or no measurable difference in charge affinity between different types of metal, possibly because the fast motion of conduction electrons cancels such differences.

Cause

Although the word comes from the Greek for "rubbing", the two materials only need to come into contact and then separate for electrons to be exchanged. After coming into contact, a chemical bond is formed between some parts of the two surfaces, called adhesion, and charges move from one material to the other to equalize their electrochemical potential. This is what creates the net charge imbalance between the objects. When separated, some of the bonded atoms have a tendency to keep extra electrons, and some a tendency to give them away, though the imbalance will be partially destroyed by tunnelling or electrical breakdown (usually corona discharge). In addition, some materials may exchange ions of differing mobility, or exchange charged fragments of larger molecules.

The triboelectric effect is related to friction only because they both involve adhesion. However, the effect is greatly enhanced by rubbing the materials together, as they touch and separate many times. For surfaces with differing geometry, rubbing may also lead to heating of protrusions, causing pyroelectric charge separation which may add to the existing contact electrification, or which may oppose the existing polarity. Surface nano-effects are not well understood, and the atomic force microscope has made sudden progress possible in this field of physics.

Because the surface of the material is now electrically charged, either negatively or positively, any contact with an uncharged conductive object or with an object having substantially different charge may cause an electrical discharge of the built-up static electricity; a spark. A person simply walking across a carpet may build up a charge of many thousands of volts, enough to cause a spark one centimeter long or more. Low relative humidity in the ambient air increases the voltage at which electrical discharge occurs by increasing the ability of the insulating material to hold charge and by decreasing the conductivity of the air, making it difficult for the charge build-up to dissipate gradually. Simply removing a nylon shirt or corset can also create sparks. Car travel can lead to a build-up of charge on the driver and passengers due to friction between the drivers clothes and the leather or plastic furnishings inside the vehicle. This charge can then be relaxed as a spark to the metal car body, fuel dispensers, or nearby door handles, etc. While the vehicle's body itself can build up a static charge (which acts as a Faraday cage) it can relax through the carbon in the tires. If it remains charged when parked, sparks may jump from the door frame to occupants as they make contact with the ground.

This type of discharge is often harmless because the energy $((V^2 * C)/2)$ of the spark is very small, being typically several tens of micro joules in cold dry weather, and much less than that in humid conditions. However, such sparks can ignite flammable vapours.

In Aircraft and Spacecraft

Aircraft flying in weather will develop a static charge from air friction on the airframe. The static can be discharged with static dischargers or static wicks.

NASA follows what they call the Triboelectrification Rule whereby they will cancel a launch if the launch vehicle is predicted to pass through certain types of clouds. Flying through high-level clouds can generate "P-static" (P for precipitation), which can create static around the launch vehicle that will interfere with radio signals sent by or to the vehicle. This may prevent transmitting of telemetry to the ground or, if the need arises, to send a signal to the vehicle, particularly any critical signal for the flight termination system. When a hold is put in place due to the triboelectrification rule, it remains until Space Wing and observer personnel such as those in reconnaissance aircraft indicate that the skies are clear.

Risks and Counter-measures

Ignition: The effect is of considerable industrial importance in terms of both safety and potential damage to manufactured goods.

Static discharge is a particular hazard in grain elevators owing to the danger of a dust explosion. The spark produced is fully able to ignite flammable vapours, for example, petrol, ether fumes as well as methane gas. For bulk fuel deliveries and aircraft fuelling a grounding connection is made between the vehicle and the receiving tank prior to opening the tanks. When fuelling vehicles at a retail station it is proper to touch metal on the car before opening the gas tank or touching the nozzle.

In the Workplace

Means have to be provided to discharge carts which may carry such volatile liquids, flammable gasses, or oxygen in hospitals. Even where only a small charge is produced, it can result in dust particles being attracted to the rubbed surface. In the case of textile manufacture this can lead to a permanent grimy mark where the cloth comes in contact with dust accumulations held by a static charge. Dust attraction may be reduced by treating insulating surfaces with an antistatic cleaning agent.

Damage to Electronics

Some electronic devices, most notably CMOS integrated circuits and MOSFET transistors, can be accidentally destroyed by high-voltage static discharge. Such components are usually stored in a conductive foam for protection. Grounding oneself by touching the workbench, others, or using a special bracelet or anklet is standard practice while handling unconnected integrated circuits. Another way of dissipating charge is by using conducting materials such as carbon black loaded rubber mats in operating theatres, for example.

Devices containing sensitive components must be protected from normal use, installation, and disconnection, accomplished by designed-in protection at external connections where needed. Protection may be through the use of more robust devices or protective countermeasures at the device's external interfaces. These may be opto-isolators, less sensitive types of transistors, and static bypass devices such as metal oxide varistors.

6

Centroid and Centre of Gravity

Centroid of Simple Figures from First Principle

In mathematics and physics, the centroid or geometric centre of a two-dimensional region is, informally, the point at which a cardboard cut-out of the region could be perfectly balanced on the tip of a pencil (assuming uniform density and a uniform gravitational field). Formally, the centroid of a plane figure or two-dimensional shape is the arithmetic mean ("average") position of all the points in the shape. The definition extends to any object in n-dimensional space: its centroid is the mean position of all the points in all of the coordinate directions.

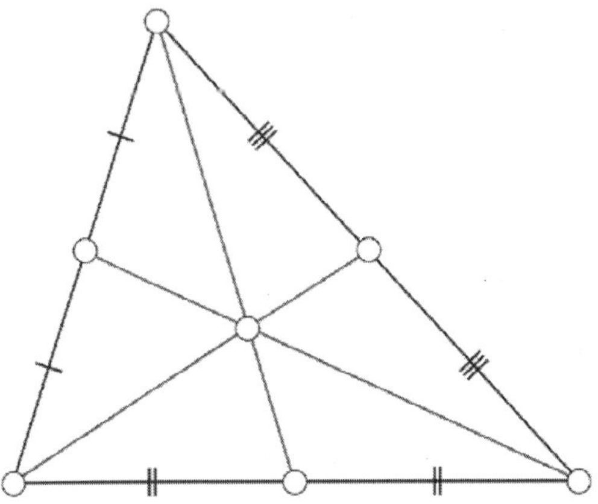

Figure: *Centroid of a triangle*

While in geometry the term barycentre is a synonym for "centroid", in physics "barycentre" may also mean the physical centre of mass or the centre of gravity, depending on the context.

The centre of mass (and centre of gravity in a uniform gravitational field) is the arithmetic mean of all points weighted by the local density or specific weight. If a physical object has uniform density, then its centre of mass is the same as the centroid of its shape. In geography, the centroid of a radial projection of a region of the Earth's surface to sea level is known as the region's geographical centre.

Properties

The geometric centroid of a convex object always lies in the object. A non-convex object might have a centroid that is outside the figure itself. The centroid of a ring or a bowl, for example, lies in the object's central void. If the centroid is defined, it is a fixed point of all isometries in its symmetry group. In particular, the geometric centroid of an object lies in the intersection of all its hyperplanes of symmetry. The centroid of many figures (regular polygon, regular polyhedron, cylinder, rectangle, rhombus, circle, sphere, ellipse, ellipsoid, superellipse, superellipsoid, etc.) can be determined by this principle alone. In particular, the centroid of a parallelogram is the meeting point of its two diagonals. This is not true for other quadrilaterals.

For the same reason, the centroid of an object with translational symmetry is undefined (or lies outside the enclosing space), because a translation has no fixed point.

Locating the Centroid

Plumb Line Method: The centroid of a uniform planar lamina, such as (a) below, may be determined, experimentally, by using a plumbline and a pin to find the centre of mass of a thin body of uniform density having the same shape. The body is held by the pin inserted at a point near the body's perimeter, in such a way that it can freely rotate around the pin; and the plumb line is dropped from the pin (b). The position of the plumbline is traced on the body. The experiment is repeated with the pin inserted at a different point of the object. The intersection of the two lines is the centroid of the figure (c).

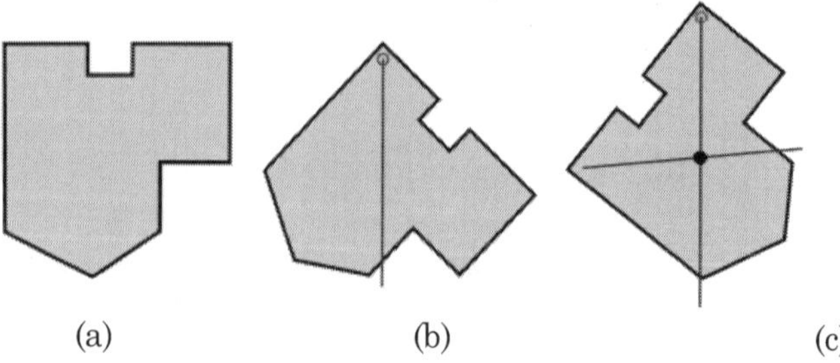

(a) (b) (c)

This method can be extended (in theory) to concave shapes where the centroid lies outside the shape, and to solids (of uniform density), but the positions of the plumb lines need to be recorded by means other than drawing.

Balancing Method

For convex two-dimensional shapes, the centroid can be found by balancing the shape on a smaller shape, such as the top of a narrow cylinder. The centroid occurs somewhere within the range of contact between the two shapes. In principle, progressively narrower cylinders can be used to find the centroid to arbitrary accuracy. In practice air currents make this unfeasible. However, by marking the overlap range from multiple balances, one can achieve a considerable level of accuracy.

Of a Finite Set of Points

The centroid of a finite set of k points $x_1, x_2, \ldots, x_k$ in $\mathbb{R}^n$ is

$$C = \frac{x_1 + x_2 + \cdots + x_k}{k}$$

This point minimizes the sum of squared Euclidean distances between itself and each point in the set.

By Integral Formula

The centroid of a subset X of $\mathbb{R}^n$ can also be computed by the integral

$$C = \frac{\int xg(x)\,dx}{\int g(x)\,dx}$$

where the integrals are taken over the whole space $\mathbb{R}^n$, and g is the characteristic function of the subset, which is 1 inside X and 0 outside it. Note that the denominator is simply the measure of the set X (However, this formula cannot be applied if the set X has zero measure, or if either integral diverges.)

Another formula for the centroid is

$$C_k = \frac{\int zS_k(z)\,dz}{\int S_k(z)\,dz}$$

where C_k is the kth coordinate of C, and $S_k(z)$ is the measure of the intersection of X with the hyperplane defined by the equation $x_k = z$. Again, the denominator is simply the measure of X.

For a plane figure, in particular, the barycentre coordinates are

$$C_x = \frac{\int x S_y(x)\, dx}{A}$$

$$C_y = \frac{\int y S_x(y)\, dy}{A}$$

where A is the area of the figure X; $S_y(x)$ is the length of the intersection of X with the vertical line at abscissa x; and $S_x(y)$ is the analogous quantity for the swapped axes.

Bounded Region

The centroid $(\bar{x}, \bar{y})$ of a region bounded by the graphs of the continuous functions f and g such that $f(x) \geq g(x)$ on the interval $[a,b]$, $a \leq x \leq b$, is given by

$$\bar{x} = \frac{1}{A}\int_a^b x[f(x) - g(x)]\, dx$$

$$\bar{y} = \frac{1}{A}\int_a^b \left[\frac{f(x) + g(x)}{2}\right][f(x) - g(x)]\, dx,$$

where A is the area of the region (given by $\int_a^b [f(x) - g(x)]dx$).

Example

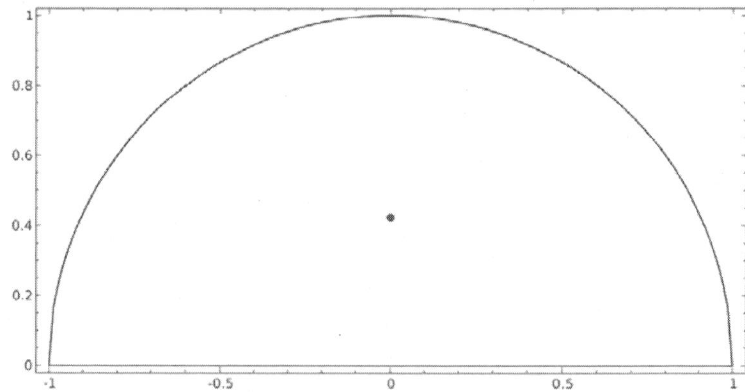

Figure: *Semicircle with a red dot showing the centroid*

Consider the semicircle bounded by $f(x) = \sqrt{1 - x^2}$ and $g(x) = 0$. Its area is $A = \frac{\pi r^2}{2} = \frac{\pi}{2}$.

$$\bar{x} = \frac{1}{A}\int_a^b x[f(x) - g(x)]\, dx = \frac{2}{\pi}\int_{-1}^{1} x\sqrt{1 - x^2}\, dx = 0$$

$$\bar{y} = \frac{1}{A}\int_a^b \left[\frac{f(x)+g(x)}{2}\right][f(x)-g(x)]\,dx = \frac{2}{\pi}\int_{-1}^{1}\left[\frac{\left(\sqrt{1-x^2}\right)^2}{2}\right]dx = \frac{4}{3\pi}$$

The centroid is located at $(0, \frac{4}{3\pi})$.

Of an L-shaped Object

This is a method of determining the centroid of an L-shaped object.

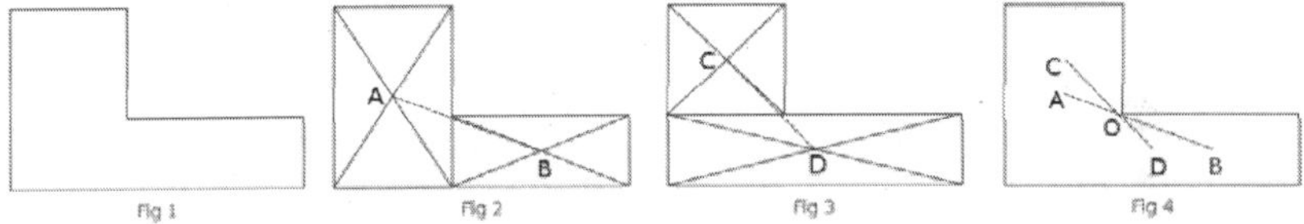

1. Divide the shape into two rectangles. Find the centroids of these two rectangles by drawing the diagonals. Draw a line joining the centroids. The centroid of the shape must lie on this line AB.
2. Divide the shape into two other rectangles. Find the centroids of these two rectangles by drawing the diagonals. Draw a line joining the centroids. The centroid of the L-shape must lie on this line CD.
3. As the centroid of the shape must lie along AB and also along CD, it is obvious that it is at the intersection of these two lines, at O. The point O might *not* lie inside the L-shaped object.

Of Triangle and Tetrahedron

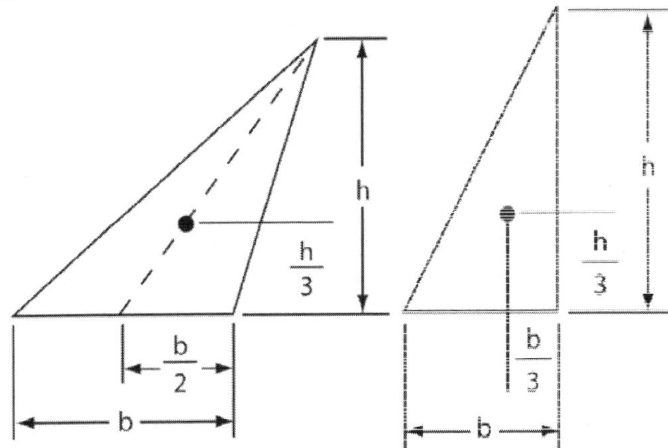

The centroid of a triangle is the point of intersection of its medians (the lines joining each vertex with the midpoint of the opposite side). The centroid divides each of the medians in the ratio 2:1, which is to say it is located S! of the perpendicular distance between each side and the opposing point. Its Cartesian coordinates are the means of the coordinates of the three vertices. That is, if the three vertices are $a = (x_a, y_a)$, $b = (x_b, y_b)$, and $c = (x_c, y_c)$, then the centroid is

$$C = \frac{1}{3}(a + b + c) = \left(\frac{1}{3}(x_a + x_b + x_c),\ \frac{1}{3}(y_a + y_b + y_c)\right).$$

The centroid is therefore at $\left(\frac{1}{3}, \frac{1}{3}, \frac{1}{3}\right)$ in barycentric coordinates.

The centroid is also the physical centre of mass if the triangle is made from a uniform sheet of material; or if all the mass is concentrated at the three vertices, and evenly divided among them. On the other hand, if the mass is distributed along the triangle's perimeter, with uniform linear density, then the centre of mass lies at the Spieker centre (the incentre of the medial triangle), which does not (in general) coincide with the geometric centroid of the full triangle.

The area of the triangle is 1.5 times the length of any side times the perpendicular distance from the side to the centroid.

A triangle's centroid lies on its Euler line between its orthocentre and its circumcentre, exactly twice as close to the latter as to the former. Similar results hold for a tetrahedron: its centroid is the intersection of all line segments that connect each vertex to the centroid of the opposite face. These line segments are divided by the centroid in the ratio 3:1. The result generalizes to any n-dimensional simplex in the obvious way. If the set of vertices of a simplex is $v_0, \ldots, v_n$, then considering the vertices as vectors, the centroid is

$$C = \frac{1}{n+1} \sum_{i=0}^{n} v_i.$$

The geometric centroid coincides with the centre of mass if the mass is uniformly distributed over the whole simplex, or concentrated at the vertices as n equal masses.

The isogonal conjugate of a triangle's centroid is its symmedian point.

Centroid of Polygon

The centroid of a non-self-intersecting closed polygon defined by n vertices (x_0,y_0), (x_1,y_1), ..., $(x_{n"1},y_{n-1})$ is the point (C_x, C_y), where

$$C_x = \frac{1}{6A} \sum_{i=0}^{n-1} (x_i + x_{i+1})(x_i\ y_{i+1} - x_{i+1}\ y_i)$$

$$C_y = \frac{1}{6A}\sum_{i=0}^{n-1}(y_i + y_{i+1})(x_i\, y_{i+1} - x_{i+1}\, y_i)$$

and where A is the polygon's signed area,

$$A = \frac{1}{2}\sum_{i=0}^{n-1}(x_i\, y_{i+1} - x_{i+1}\, y_i)$$

In these formulas, the vertices are assumed to be numbered in order of their occurrence along the polygon's perimeter, and the vertex (x_n, y_n) is assumed to be the same as (x_0, y_0). Note that if the points are numbered in clockwise order the area A, computed as above, will have a negative sign; but the centroid coordinates will be correct even in this case.

Centroid of Cone or Pyramid

The centroid of a cone or pyramid is located on the line segment that connects the apex to the centroid of the base. For a solid cone or pyramid, the centroid is 1/4 the distance from the base to the apex. For a cone or pyramid that is just a shell (hollow) with no base, the centroid is 1/3 the distance from the base plane to the apex.

Centre of Gravity

Centre of gravity is the point in a body around which the resultant torque due to gravity forces vanish. Near the surface of the earth, where the gravity acts downward as a parallel force field, the centre of gravity and the centre of mass are the same.

The study of the dynamics of aircraft, vehicles and vessels assumes that the system moves in near-earth gravity, and therefore the terms centre of gravity and centre of mass are used interchangeably.

In physics the benefits of using the centre of mass to model a mass distribution can be seen by considering the resultant of the gravity forces on a continuous body. Consider a body of volume V with density ρ(r) at each point r in the volume. In a parallel gravity field the force f at each point r is given by,

$$\mathrm{f(r)} = -dm\,g\vec{k} = -\rho(\mathrm{r})dV\,g\vec{k},$$

where dm is the mass at the point r, g is the acceleration of gravity, and k is a unit vector defining the vertical direction. Choose a reference point R in the volume and compute the resultant force and torque at this point,

$$\mathrm{F} = \int_V \mathrm{f(r)} = \int_V \rho(\mathrm{r})dV(-g\vec{k}) = -Mg\vec{k},$$

and

$$T = \int_V (r - R) \times f(r) = \int_V (r - R) \times (-g\rho(r)dV\vec{k}) = \left(\int_V \rho(r)(r - R)dV\right) \times (-g\vec{k}).$$

If the reference point R is chosen so that it is the centre of mass, then

$$\int_V \rho(r)(r - R)dV = 0,$$

which means the resultant torque T=0. Because the resultant torque is zero the body will move as though it is a particle with its mass concentrated at the centre of mass.

By selecting the centre of gravity as the reference point for a rigid body, the gravity forces will not cause the body to rotate, which means weight of the body can be considered to be concentrated at the centre of mass.

Area Moment of Inertia

Moment of inertia is a property of a body that defines its resistance to a change in angular velocity about an axis of rotation. It is how rotation of a body is affected by Newton's law of inertia, which states "Every body perseveres in its state of being at rest or of moving uniformly straight forward except insofar as it is compelled to change its state by forces impressed." In this context, inertia refers to resistance to change.

Moment of inertia applies to an extended body in which the mass is constrained to rotate around an axis. It arises as a combination of mass and geometry in the study of the movement of continuous bodies, or assemblies of particles, known as rigid body dynamics. It is the moment of inertia of the pole carried by a tight-rope walker that resists rotation and helps the walker maintain balance.

When a body is rotating around an axis, a torque must be applied to change its angular velocity. The amount of torque needed for any given change in angular velocity is proportional to the size of that change. The constant of proportionality is a property of the body that combines its mass and its shape, known as the moment of inertia. In classical mechanics, moment of inertia may also be called mass moment of inertia, rotational inertia, polar moment of inertia, or the angular mass (SI units kg·m^2, US units lb$_m$ ft^2).

In 1673 Christiaan Huygens introduced this parameter in his study of the oscillation of a body hanging from a pivot, known as a compound pendulum. The term *moment of inertia* was introduced by

Leonhard Euler in his book *Theoria motus corporum solidorum seu rigidorum* in 1765, and it is incorporated into Euler's second law.

The natural frequency of oscillation of a compound pendulum is obtained from the ratio of the torque imposed by gravity on the mass of the pendulum to the resistance to acceleration defined by the moment of inertia. Comparison of this natural frequency to that of a simple pendulum consisting of a single point of mass provides a mathematical formulation for moment of inertia of an extended body.

Moment of inertia also appears in momentum, kinetic energy, and in Newtons laws of motion for a rigid body as a physical parameter that combines its shape and mass. There is an interesting difference in the way moment of inertia appears in planar and spatial movement. Planar movement has a single scalar that defines the moment of inertia, while for spatial movement the same calculations yield a 3x3 matrix of moments of inertia, called the inertia matrix or inertia tensor.

The moment of inertia of a rotating flywheel is used in a machine to resist variations in applied torque in order to smooth its rotational output. The moment of inertia of an airplane about its longitudinal, horizontal and vertical axes determines how steering forces on the control surfaces of its wings, elevators and tail affect the plane in roll, pitch and yaw.

Moment of Inertia of a Pendulum

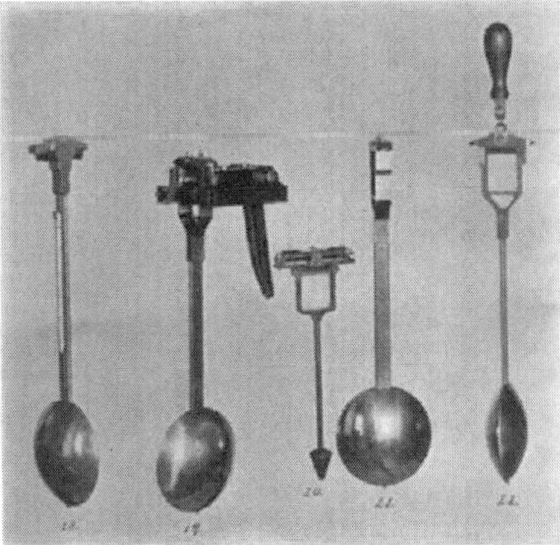

Figure: *Pendulums used in Mendenhall gravimeter apparatus, from 1897 scientific journal. The portable gravimeter developed in 1890 by Thomas C. Mendenhall provided the most accurate relative measurements of the local gravitational field of the Earth.*

The mass of a simple pendulum supported by a light string accelerates due to the force of gravity. Its movement is constrained to a circle around the pivot point, so it can be considered to be a compound pendulum that consists of a single mass point. The ratio of the torque due to gravity about the pivot and the resistance of mass to angular acceleration, the moment of inertia of the particle is found to be the product the mass of the particle with the square of its distance to the pivot. The moments of inertia of an assembly of particles, or continuous body, that rotates around the pivot is the sum the moment of inertia of each of the particles. The force of gravity on the mass of a simple pendulum generates a torque $\tau = \mathrm{r} \times \mathrm{F}$ around the axis perpendicular to the plane of the pendulum movement. Associated with this torque is an angular acceleration α of the string and mass around this axis. Let F be the component of force in the direction of the acceleration of the pendulum mass and r the length of the string, so the acceleration of the mass is $a = r\alpha$, then $F = ma$, becomes,

$$\tau = rF\vec{k} = (mr^2)\alpha = I\alpha\vec{k},$$

where $\vec{k}$ is a unit vector perpendicular to the plane of the pendulum. The quantity $I = mr^2$ is the *moment of inertia* of this single mass around the pivot point.

The quantity $I = mr^2$ also appears in the angular momentum of a simple pendulum, which is calculated from the velocity $v = r\omega$ of the pendulum mass around the pivot. This is given by

$$\mathrm{L} = \mathrm{r} \times (m\mathrm{v}) = (mr^2)\omega = I\omega\vec{k}.$$

Similarly, the kinetic energy of the pendulum mass is defined by the velocity of the pendulum around the pivot to yield

$$E_\mathrm{K} = \frac{1}{2} m\mathrm{v} \cdot \mathrm{v} = \frac{1}{2}(mr^2)\omega^2 = \frac{1}{2} I\omega^2.$$

This shows that the quantity $I = mr^2$ is how mass combines with the shape of a body to define rotational inertia. The moment of inertia of an arbitrarily shaped body is the sum of the values mr^2 for all of the elements of mass in the body.

Compound Pendulum

A compound pendulum is a body formed from an assembly of particles or continuous shapes, and its moment of inertia is the sum of the moments of inertia of all of the components in this assembly about the pivot point. The natural frequency (ω_n) of a compound pendulum depends on its moment of inertia, I_P,

$$\omega_n = \sqrt{\frac{mgr}{I_P}},$$

where m is the mass of the object, g is local acceleration of gravity, and r is the distance from the pivot point to the centre of mass of the object. Measuring this frequency of oscillation over small angular displacements provides an effective way of measuring moment of inertia of a body.

Thus, to determine the moment of inertia of the body, simply suspend it from a convenient pivot point $_P$ so that it swings freely in a plane perpendicular to the direction of the desired moment of inertia, then measure its natural frequency or period of oscillation (t), to obtain

$$I_P = \frac{mgr}{\omega_n^2} = \frac{mgrt^2}{4\pi^2},$$

where t is the period (duration) of oscillation (usually averaged over multiple periods). The moment of inertia of the body about its centre of mass, I_C, is then calculated using the parallel axis theorem to be

$$I_C = I_P - mr^2,$$

where m is the mass of the body and r is the distance from the pivot point p to the centre of mass c.

Moment of inertia of a body is often defined in terms of its *radius of gyration*, which is the radius of a ring of equal mass around the centre of mass of a body that has the same moment of inertia. The radius of gyration K is calculated from the body's moment of inertia I_C and mass as the length,

$$K = \sqrt{\frac{I_C}{m}}.$$

Centre of Oscillation

A simple pendulum that has the same natural frequency as a compound pendulum defines the length L from the pivot to a point called the centre of oscillation of the compound pendulum. This point also corresponds to the centre of percussion. The length L is determined from the formula,

$$\omega_n = \sqrt{\frac{g}{L}} = \sqrt{\frac{mgr}{I_P}},$$

or

$$L = \frac{g}{\omega_n^2} = \frac{I_P}{mr}.$$

The seconds pendulum, which provides the "tick" and "tock" of a grandfather clock, takes one second to swing from side-to-side. This is a period of two seconds, or a natural frequency of π radians/second for the pendulum. In this case, the length L is given by,

$$L = \frac{g}{\omega_n^2} = \frac{9.81 \text{ m/s}^2}{(3.14 \text{ rad/s})^2} = 0.99 \text{ m}.$$

Notice that the centre of oscillation of the seconds pendulum must be adjusted to accommodate use in locations with different values for the local acceleration of gravity. Kater's pendulum is an example of a compound pendulum that is used to measure gravity called a gravimeter.

Measuring Moment of Inertia

The moment of inertia of complex systems such as a vehicle or airplane around its vertical axis can be measured by suspending the system from three points to form a trifilar pendulum. A trifilar pendulum is a platform supported by three wires designed to oscillate in torsion around its vertical centroidal axis. The period of oscillation of the trifilar pendulum yields the moment of inertia of the system.

Calculating Moment of Inertia

Figure: *Four objects racing down a plane while rolling without slipping. From back to front: spherical shell (red), solid sphere (orange), cylindrical ring (green) and solid cylinder (blue). The time for each object to reach the finishing line depends on their moment of inertia. (Details, Animated GIF version)*

The moment of inertia of a body is calculated by summing mr^2 for every particle in the body about a given rotation axis. In order to see how moment of inertia arises in the study of the movement of an extended body, it is convenient to consider a rigid assembly of point masses.

Consider the kinetic energy of an assembly of N masses m_i that lie at the distances r_i from the pivot point P, which is the sum of the kinetic energy of the individual masses,

$$E_K = \sum_{i=1}^{N} \frac{1}{2} m_i \mathrm{v}_i \cdot \mathrm{v}_i = \sum_{i=1}^{N} \frac{1}{2} m_i (\omega r_i)^2 = \frac{1}{2} \omega^2 \sum_{i=1}^{N} m_i r_i^2.$$

This shows that the moment of inertia of the body is the sum of each of the mr^2 terms, that is

$$I_P = \sum_{i=1}^{N} m_i r_i^2.$$

Thus, moment of inertia is a physical property that combines the mass and distribution of the particles around the rotation axis. Notice that rotation about different axes of the same body yield different moments of inertia.

The moment of inertia of a continuous body rotating about a specified axis is calculated in the same way, with the summation replaced by the integral,

$$I_P = \int_V \rho(\mathrm{r}) \mathrm{r}^2 \, dV.$$

Again r is the radius vector to a point in the body from the specified axis through the pivot P, and $\rho(\mathrm{r})$ is the mass density at each point r. The integration is evaluated over the volume V of the body.

Note on second moment of area: The moment of inertia of a body moving in a plane and the second moment of area of a beam's cross-section are often confused. A beam along the z-axis has stresses in the cross-section in the x-y plane that are calculated using the second moment of this area around either the x-axis or y-axis depending on the load. The moment of inertia of mass distributed along a body with the shape of this cross-section is the second moment of this area about the z-axis weighted by its density. The second moment of area around an axis perpendicular to the area is called the *polar moment of the area*, and is the sum of the second moments about the x and y axes.

Example Calculation of Moment of Inertia

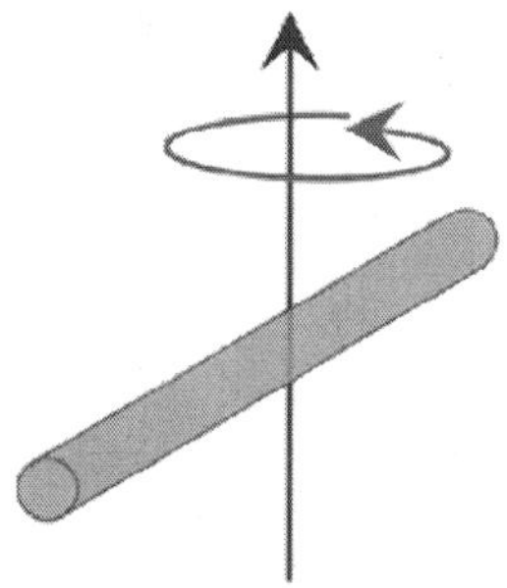

The moment of inertia of a compound pendulum constructed from a thin disc mounted at the end of a thin rod that oscillates around a pivot at the other end of the rod, begins with the calculation of the moment of inertia of the thin rod and thin disc about their respective centres of mass.

- The moment of inertia of a thin rod with constant cross-section s and density ρ and with length l about a perpendicular axis through its centre of mass is determined by integration. Align the x-axis with the rod and locate the origin its centre of mass at the centre of the rod, then

$$I_{C,\text{rod}} = \int \rho x^2 dV = \int_{-\ell/2}^{\ell/2} \rho x^2 s dx = \rho s \frac{x^3}{3}\Big|_{-\ell/2}^{\ell/2} = \frac{\rho s}{3}(\ell^3/8 + \ell^3/8) = \frac{1}{12} m\ell^2,$$

where $m = \rho s \ell$ is the mass of the rod.

- The moment of an inertia of a thin disc of constant thickness s, radius R, and density ρ about an axis through its centre and perpendicular to its face (parallel to its axis of rotational symmetry) is determined by integration. Align the z-axis with the axis of the disc and define a volume element as $dV = sr\, drd\theta$, then

$$I_{C,\text{disc}} = \int \rho r^2 dV = \int_0^{2\pi} \int_0^R \rho r^2 (srdrd\theta) = 2\pi\rho s \frac{R^4}{4} = \frac{1}{2} mR^2,$$

where $m = \pi R^2 \rho s$ is its mass.

- The moment of inertia of the compound pendulum is now obtained by adding the moment of inertia of the rod and the disc around the pivot point P as,

$$I_P = I_{C,\text{rod}} + M_{\text{rod}}(L/2)^2 + I_{C,\text{disc}} + M_{\text{disc}}(L+R)^2,$$

where L is the length of the pendulum. Notice that the parallel axis theorem is used to shift the moment of inertia from the centre of mass to the pivot point of the pendulum.

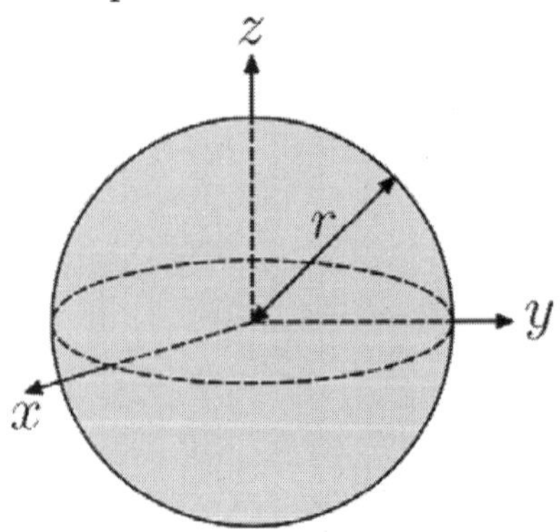

A list of moments of inertia formulas for standard body shapes provides a way to obtain the moment of inertial of a complex body as an assembly of simpler shaped bodies.

The parallel axis theorem is used to shift the reference point of the individual bodies to the reference point of the assembly.

As one more example, consider the moment of inertia of a solid sphere of constant density about an axis through its centre of mass. This is determined by summing the moment of inertias of the thin discs that form the sphere. If the surface of the ball is defined by the equation

$$x^2 + y^2 + z^2 = R^2,$$

then the radius r of the disc at the cross-section z along the z-axis is

$$r(z)^2 = x^2 + y^2 = R^2 - z^2.$$

Therefore, the moment of inertia of the ball is the sum of the moment of inertias of the discs along the z-axis,

$$I_{C,\text{ball}} = \int_{-R}^{R} \frac{\pi\rho}{2} r(z)^4\, dz = \int_{-R}^{R} \frac{\pi\rho}{2}(R^2 - z^2)^2\, dz = \frac{\pi\rho}{2}(R^4 z - 2R^2 z^3/3 + z^5/5)\Big|_{-R}^{R}$$

$$= \pi\rho(1 - 2/3 + 1/5)R^5 = \frac{2}{5} mR^2,$$

where $m = (4/3)\pi R^3\rho$ is the mass of the ball.

Moment of Inertia in Planar Movement of a Rigid Body

If a mechanical system is constrained to move parallel to a fixed plane, then the rotation of a body in the system occurs around an axis k perpendicular to this plane. In this case, the moment of inertia of the mass in this system is a scalar known as the *polar moment of inertia.* The definition of the polar moment of inertia can be obtained by considering momentum, kinetic energy and Newton's laws for the planar movement of a rigid system of particles.

If a system of n particles, P_i, $i = 1,\ldots,n$, are assembled into a rigid body, then the momentum of the system can be written in terms of positions relative to a reference point R, and absolute velocities v_i

$$\Delta r_i = r_i - R, \quad v_i = \omega \times (r_i - R) + V,$$

where ω is the angular velocity of the system and V is the velocity of R.

For planar movement the angular velocity vector is directed along the unit vector k which is perpendicular to the plane of movement. Introduce the unit vectors e_i from the reference point R to a point r_i, and the unit vector $t_i = k \times e_i$ so

$$\Delta r_i \mathrm{e}_i = \mathrm{r}_i - \mathrm{R}, \quad \mathrm{v}_i = \omega \Delta r_i \mathrm{t}_i + \mathrm{V}, \quad i = 1, \ldots, n.$$

This defines the relative position vector and the velocity vector for the rigid system of the particles moving in a plane.

Note on the cross product: When a body moves parallel to a ground plane, the trajectories of all the points in the body lie in planes parallel to this ground plane. This means that any rotation that the body undergoes must be around an axis perpendicular to this plane. Planar movement is often presented as projected onto this ground plane so that the axis of rotation appears as a point. In this case, the angular velocity and angular acceleration of the body are scalars and the fact that they are vectors along the rotation axis is ignored. This is usually preferred for introductions to the topic. But in the case of moment of inertia, the combination of mass and geometry benefits from the geometric properties of the cross product. For this reason, in this section on planar movement the angular velocity and accelerations of the body are vectors perpendicular to the ground plane, and the cross product operations are the same as used for the study of spatial rigid body movement.

Angular Momentum in Planar Movement

The angular momentum vector for the planar movement of a rigid system of particles is given by

$$\begin{aligned} \mathrm{L} &= \sum_{i=1}^{n} m_i (\mathrm{r}_i - \mathrm{R}) \times \mathrm{v}_i \\ &= \sum_{i=1}^{n} m_i \Delta r_i \mathrm{e}_i \times (\omega \Delta r_i \mathrm{t}_i + \mathrm{V}) \\ &= \left(\sum_{i=1}^{n} m_i \Delta r_i^2\right) \omega \vec{k} + \left(\sum_{i=1}^{n} m_i \Delta r_i \mathrm{e}_i\right) \times \mathrm{V}. \end{aligned}$$

Use the centre of mass C as the reference point so

$$\Delta r_i \mathrm{e}_i = \mathrm{r}_i - \mathrm{C}, \quad \sum_{i=1}^{n} m_i \Delta r_i \mathrm{e}_i = 0,$$

and define the moment of inertia relative to the centre of mass $\mathrm{I_C}$ as

$$I_C = \sum_{i=1}^{n} m_i \Delta r_i^2,$$

then the equation for angular momentum simplifies to

$$\mathrm{L} = I_C \omega \vec{k}.$$

The moment of inertia I_C about an axis perpendicular to the movement of the rigid system and through the centre of mass is known as the *polar moment of inertia.*

For a given amount of angular momentum, a decrease in the moment of inertia results in an increase in the angular velocity. Figure skaters can change their moment of inertia by pulling in their arms. Thus, the angular velocity achieved by a skater with outstretched arms results in a greater angular velocity when the arms are pulled in, because of the reduced moment of inertia.

Kinetic Energy in Planar Movement

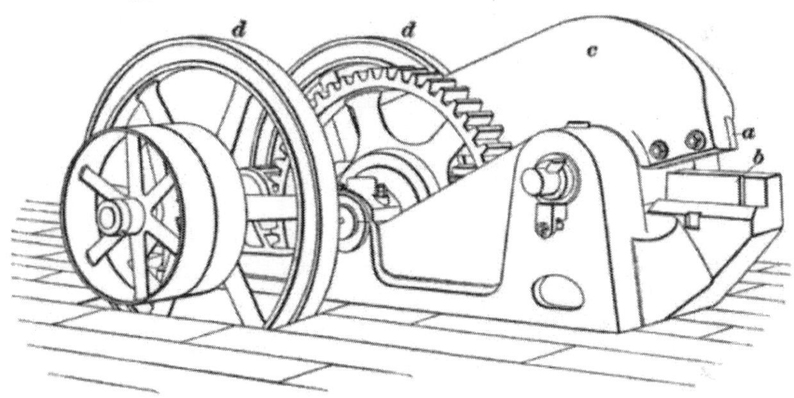

Figure: *This 1906 rotary shear uses the moment of inertia of two flywheels to store kinetic energy which when released is used to cut metal stock (International Library of Technology, 1906).*

The kinetic energy of a rigid system of particles moving in the plane is given by

$$E_K = \frac{1}{2}\sum_{i=1}^{n} m_i \mathrm{v}_i \cdot \mathrm{v}_i = \frac{1}{2}\sum_{i=1}^{n} m_i (\omega \Delta r_i \mathrm{t}_i + \mathrm{V}) \cdot (\omega \Delta r_i \mathrm{t}_i + \mathrm{V}).$$

This equation expands to yield three terms

$$E_K = \frac{1}{2}\omega^2 \sum_{i=1}^{n} m_i \Delta r_i^2 (\mathrm{t}_i \cdot \mathrm{t}_i) + \omega \mathrm{V} \cdot (\sum_{i=1}^{n} m_i \Delta r_i \mathrm{t}_i) + \frac{1}{2}(\sum_{i=1}^{n} m_i) \mathrm{V} \cdot \mathrm{V}.$$

Let the reference point be the centre of mass C of the system so the second term becomes zero, and introduce the moment of inertia I_C so the kinetic energy is given by

$$E_K = \frac{1}{2} I_C \omega^2 + \frac{1}{2} M\mathrm{V} \cdot \mathrm{V}.$$

The moment of inertia I_C is the *polar moment of inertia* of the body.

Newton's Laws for Planar Movement

Newton's laws for a rigid system of N particles, P_i, $i = 1,..., N$, can be written in terms of a resultant force and torque at a reference point R, to yield

$$F = \sum_{i=1}^{N} m_i A_i, \quad T = \sum_{i=1}^{N} (r_i - R) \times (m_i A_i),$$

where r_i denotes the trajectory of each particle.

The kinematics of a rigid body yields the formula for the acceleration of the particle P_i in terms of the position R and acceleration A of the reference particle as well as the angular velocity vector ω and angular acceleration vector α of the rigid system of particles as,

$$A_i = \alpha \times (r_i - R) + \omega \times \omega \times (r_i - R) + A.$$

For systems that are constrained to planar movement, the angular velocity and angular acceleration vectors are directed along k perpendicular to the plane of movement, which simplifies this acceleration equation. In this case, the acceleration vectors can be simplified by introducing the unit vectors e_i from the reference point R to a point r_i and the unit vectors $t_i = k \times e_i$, so

$$A_i = \alpha(\Delta r_i t_i) - \omega^2 (\Delta r_i e_i) + A.$$

This yields the resultant torque on the system as

$$\tau = \sum_{i=1}^{N} (m_i \Delta r_i e_i) \times (\alpha(\Delta r_i t_i) - \omega^2 (\Delta r_i e_i) + A) = (\sum_{i=1}^{N} m_i \Delta r_i^2) \alpha \vec{k} + (\sum_{i=1}^{N} m_i \Delta r_i e_i) \times A,$$

where $e_i \times e_i = 0$, and $e_i \times t_i = k$ is the unit vector perpendicular to the plane for all of the particles P_i.

Use the centre of mass C as the reference point and define the moment of inertia relative to the centre of mass I_C, then the equation for the resultant torque simplifies to

$$\tau = I_C \alpha \vec{k}.$$

The parameter I_C is the *polar moment of inertia* of the moving body.

The Inertia Matrix for Spatial Movement of a Rigid Body

The scalar moments of inertia appear as elements in a matrix when a system of particles is assembled into a rigid body that moves in three dimensional space. This inertia matrix appears in the calculation of the angular momentum, kinetic energy and resultant torque of the rigid system of particles.

An important application of the inertia matrix and Newton's laws of motion is the analysis of a spinning top. This is discussed in the article on Gyroscopic precession. A more detailed presentation can be found in the article on Euler's equations of motion.

Let the system of particles P_i, $i = 1,..., n$ be located at the coordinates r_i with velocities v_i relative to a fixed reference frame. For a (possibly moving) reference point R, the relative positions are

$$\Delta r_i = r_i - R$$

and the (absolute) velocities are

$$v_i = \omega \times \Delta r_i + V_R$$

where ω is the angular velocity of the system, and V_R is the velocity of R.

Angular Momentum

If the reference point R in the assembly, or body, is chosen as the centre of mass C, then its angular momentum takes the form,

$$L = \sum_{i=1}^{n} m_i \Delta r_i \times v_i = \sum_{i=1}^{n} m_i \Delta r_i \times (\omega \times \Delta r_i),$$

where the terms containing V_R sum to zero by definition of the centre of mass.

In order to define the inertia matrix, introduce the skew-symmetric matrix $[B]$ constructed from a vector b that performs the cross product operation, such that

$$[B]y = b \times y.$$

This matrix $[B]$ has the components of $b = (b_x, b_y, b_z)$ as its elements, in the form

$$[B] = \begin{bmatrix} 0 & -b_z & b_y \\ b_z & 0 & -b_x \\ -b_y & b_x & 0 \end{bmatrix}.$$

Now construct the skew-symmetric matrix $[\Delta r_i] = [r_i - C]$ obtained from the relative position vector $\Delta r_i = r_i - C$, and use this skew-symmetric matrix to define,

$$L = (-\sum_{i=1}^{n} m_i [\Delta r_i]^2)\omega = [I_C]\omega,$$

where $[I_C]$ defined by

$$[I_C] = -\sum_{i=1}^{n} m_i[\Delta r_i]^2,$$

is the inertia matrix of the rigid system of particles measured relative to the centre of mass *C*.

Kinetic Energy

The kinetic energy of a rigid system of particles can be formulated in terms of the centre of mass and a matrix of mass moments of inertia of the system. Let the system of particles P_i, $i = 1,...,n$ be located at the coordinates r_i with velocities v_i, then the kinetic energy is

$$E_K = \frac{1}{2}\sum_{i=1}^{n} m_i v_i \cdot v_i = \frac{1}{2}\sum_{i=1}^{n} m_i(\omega \times \Delta r_i + V_C)\cdot(\omega \times \Delta r_i + V_C),$$

where $\Delta r_i = r_i - C$ is the position vector of a particle relative to the centre of mass.

This equation expands to yield three terms

$$E_K = \frac{1}{2}\sum_{i=1}^{n} m_i(\omega \times \Delta r_i)\cdot(\omega \times \Delta r_i)) + \sum_{i=1}^{n} m_i V_C\cdot(\omega \times \Delta r_i)) + \frac{1}{2}\sum_{i=1}^{n} m_i V_C \cdot V_C.$$

The second term in this equation is zero because C is the centre of mass. Introduce the skew-symmetric matrix $[\Delta r_i]$ so the kinetic energy becomes

$$E_K = \frac{1}{2}\omega\cdot(-\sum_{i=1}^{n} m_i[\Delta r_i]^2)\omega + \frac{1}{2}(\sum_{i=1}^{n} m_i)V_C^2.$$

Thus, the kinetic energy of the rigid system of particles is given by

$$E_K = \frac{1}{2}\omega\cdot[I_C]\omega + \frac{1}{2}MV_C^2.$$

where $[I_C]$ is the inertia matrix relative to the centre of mass and M is the total mass.

Resultant Torque

The inertia matrix appears in the application of Newton's second law to a rigid assembly of particles. The resultant torque on this system is,

$$\tau = \sum_{i=1}^{n} (r_i - R) \times (m_i a_i),$$

where a_i is the acceleration of the particle P_i. The kinematics of a rigid body yields the formula for the acceleration of the particle P_i in terms of the position R and acceleration A of the reference point, as well as the angular velocity vector ω and angular acceleration vector α of the rigid system as,

$$a_i = \alpha \times (r_i - R) + \omega \times \omega \times (r_i - R) + A_R.$$

Use the centre of mass C as the reference point, and introduce the skew-symmetric matrix $[\Delta r_i]=[r_i - C]$ to represent the cross product $(r_i - C)x$, in order to obtain

$$\tau = (-\sum_{i=1}^{n} m_i[\Delta r_i]^2) \times \alpha + \omega \times (-\sum_{i=1}^{n} m_i[\Delta r_i]^2)\omega.$$

This calculation uses the identity

$$\Delta r_i \times (\omega \times (\omega \times \Delta r_i)) + \omega \times (\Delta r_i \times (\Delta r_i \times \omega)) = 0,$$

obtained from the Jacobi identity for the triple cross product.

Thus, the resultant torque on the rigid system of particles is given by

$$\tau = [I_C]\alpha + \omega \times [I_C]\omega,$$

where $[I_C]$ is the inertia matrix relative to the centre of mass.

Parallel Axis Theorem

The inertia matrix of a body depends on the choice of the reference point. There is a useful relationship between the inertia matrix relative to the centre of mass C and the inertia matrix relative to another point R. This relationship is called the parallel axis theorem.

Consider the inertia matrix $[I_R]$ obtained for a rigid system of particles measured relative to a reference point R, given by

$$[I_R] = -\sum_{i=1}^{n} m_i[r_i - R]^2.$$

Let C be the centre of mass of the rigid system, then

$$R = (R - C) + C = d + C,$$

where d is the vector from the centre of mass C to the reference point R. Use this equation to compute the inertia matrix,

$$[I_R] = -\sum_{i=1}^{n} m_i[r_i - C - d]^2.$$

Expand this equation to obtain

$$[I_R] = (-\sum_{i=1}^{n} m_i[r_i - C]^2) + (\sum_{i=1}^{n} m_i[r_i - C])[d] + [d](\sum_{i=1}^{n} m_i[r_i - C]) + (-\sum_{i=1}^{n} m_i)[d][d].$$

The first term is the inertia matrix $[I_C]$ relative to the centre of mass. The second and third terms are zero by definition of the centre of mass C. And the last term is the total mass of the system multiplied by the square of the skew-symmetric matrix $[d]$ constructed from d.

The result is the parallel axis theorem,

$$[I_R]=[I_C]-M[d]^2,$$

where d is the vector from the centre of mass C to the reference point R.

Note on the minus sign: By using the skew symmetric matrix of position vectors relative to the reference point, the inertia matrix of each particle has the form $-m[\mathrm{r}]^2$, which is similar to the mr^2 that appears in planar movement. However, to make this to work out correctly a minus sign is needed. This minus sign can be absorbed into the term $m[\mathrm{r}]^{\mathrm{T}}[\mathrm{r}]$, if desired, by using the skew-symmetry property of [r].

The Inertia Matrix and the Scalar Moment of Inertia around an Arbitrary Axis

The relationship between the inertia matrix of a rigid body and the scalar moment of inertia of the same body about a specified axis is important and rarely presented in detail. The following calculation expands the derivation presented by Kane and Levinson.

Let a rigid assembly of rigid system of N particles, P_i, $i = 1,...,N$, have coordinates r_i. Choose R as a reference point and compute the moment of inertia around an axis L defined by the unit vector S through the reference point R. The moment of inertia of the system around this line L=R+tS is computed by determining the perpendicular vector from this axis to the particle P_i given by

$$\Delta\mathrm{r}_i^{\perp}=(\mathrm{r}_i-\mathrm{R})-(\mathrm{S}\cdot(\mathrm{r}_i-\mathrm{R}))\mathrm{S}=[[I]-[\mathrm{SS}^T]](\Delta\mathrm{r}_i),$$

where [I] is the identity matrix and $[\mathrm{S\,S}^{\mathrm{T}}]$ is the outer product matrix formed from the unit vector S along the line L.

In order to relate this scalar moment of inertia to the inertia matrix of the body, introduce the skew-symmetric matrix [S] such that [S]y=S x y, then we have the identity

$$-[S]^2=[I]-[\mathrm{SS}^T],$$

which relies on the fact that S is a unit vector.

The magnitude squared of the perpendicular vector is

$$|\Delta\mathrm{r}_i^{\perp}|^2=(-[S]^2(\Delta\mathrm{r}_i))\cdot(-[S]^2(\Delta\mathrm{r}_i))=-\mathrm{S}\cdot[\Delta r_i][\Delta r_i]\mathrm{S}.$$

The simplification of this equation uses the identity

$$(\mathrm{S}\times(\mathrm{S}\times(\Delta \mathrm{r}_i)))\cdot\mathrm{S}\times(\mathrm{S}\times(\Delta \mathrm{r}_i)) = (\mathrm{S}\times(\mathrm{S}\times(\Delta \mathrm{r}_i)))\times\mathrm{S}\cdot(\mathrm{S}\times(\Delta \mathrm{r}_i)),$$

where the dot and the cross products have been interchanged. Expand the cross products to compute

$$-(\Delta \mathrm{r}_i)\times\mathrm{S}\cdot(\mathrm{S}\times(\Delta \mathrm{r}_i) = -\mathrm{S}\cdot[\Delta r_i][\Delta r_i]\mathrm{S},$$

where $[\Delta r_i]$ is the skew symmetric matrix obtained from the vector $\Delta r = r_i$-R.

Thus, the moment of inertia around the line L through R in the direction S is obtained from the calculation

$$I_L = \sum_{i=1}^{N} m_i\,|\Delta \mathrm{r}_i^{\perp}|^2 = -\sum_{i=1}^{N} m_i \mathrm{S}\cdot[\Delta r_i]^2\mathrm{S},$$

or

$$I_L = \mathrm{S}\cdot(-\sum_{i=1}^{N} m_i[\Delta r_i]^2)\mathrm{S} = \mathrm{S}\cdot[I_R]\mathrm{S} = \mathrm{S}^T[I_R]\mathrm{S},$$

where $[I_R]$ is the moment of inertia matrix of the system relative to the reference point R.

This shows that the inertia matrix can be used to calculate the moment of inertia of a body around any specified rotation axis in the body.and we also know that inertia can provide some moment of force

The Inertia Tensor

The inertia matrix is often described as the inertia tensor, which consists of the same moments of inertia and products of inertia about the three coordinate axes. The inertia tensor is constructed from the nine component tensors, (the symbol $\otimes$ is the tensor product)

$$\mathrm{e}_i \otimes \mathrm{e}_j, \quad i, j = 1, 2, 3,$$

where e_i, i=1,2,3 are the three orthogonal unit vectors defining the inertial frame in which the body moves. Using this basis the inertia tensor is given by

$$\mathrm{I} = \sum_{i=1}^{3}\sum_{j=1}^{3} I_{ij}\mathrm{e}_i \otimes \mathrm{e}_j.$$

This tensor is of degree two because the component tensors are each constructed from two basis vectors. In this form the inertia tensor is also called the *inertia binor*. For a rigid system of particles P_k, $k = 1,\ldots,N$ each of mass m_k with position coordinates $r_k=(x_k, y_k, z_k)$, the inertia tensor is given by

$$\mathrm{I} = \sum_{k=1}^{N} m_k ((\mathrm{r}_k \cdot \mathrm{r}_k)\mathrm{E} - \mathrm{r}_k \otimes \mathrm{r}_k),$$

where E is the identity tensor

$$\mathrm{E} = \mathrm{e}_1 \otimes \mathrm{e}_1 + \mathrm{e}_2 \otimes \mathrm{e}_2 + \mathrm{e}_3 \otimes \mathrm{e}_3.$$

The inertia tensor for a continuous body is given by

$$\mathrm{I} = \int_V \rho(\mathrm{r})\big((\mathrm{r}\cdot\mathrm{r})\mathrm{E} - \mathrm{r} \otimes \mathrm{r}\big) dV,$$

where r defines the coordinates of a point in the body and ρ(r) is the mass density at that point. The integral is taken over the volume *V* of the body. The inertia tensor is symmetric because $\mathrm{I}_{ij} = \mathrm{I}_{ji}$.

The inertia tensor can be used in the same way as the inertia matrix to compute the scalar moment of inertia about an arbitrary axis in the direction n,

$$I_n = \mathrm{n}\cdot\mathrm{I}\cdot\mathrm{n},$$

where the dot product is taken with the corresponding elements in the component tensors. A product of inertia term such as I_{12} is obtained by the computation

$$I_{12} = \mathrm{e}_1 \cdot \mathrm{I} \cdot \mathrm{e}_2,$$

and can be interpreted as the moment of inertia around the x-axis when the object rotates around the y-axis.

The components of tensors of degree two can be assembled into a matrix. For the inertia tensor this matrix is given by,

$$[I] = \begin{bmatrix} I_{11} & I_{12} & I_{13} \\ I_{21} & I_{22} & I_{23} \\ I_{31} & I_{32} & I_{33} \end{bmatrix} = \begin{bmatrix} I_{xx} & I_{xy} & I_{xz} \\ I_{xy} & I_{yy} & I_{yz} \\ I_{xz} & I_{yz} & I_{zz} \end{bmatrix}.$$

It is common in rigid body mechanics to use notation that explicitly identifies the x, y, and z axes, such as I_{xx} and I_{xy}, for the components of the inertia tensor.

Identities for a Skew-symmetric Matrix

In order to compute moment of inertia of a mass around an axis, the perpendicular vector from the mass to the axis is needed. If the axis L is defined by the unit vector S through the reference point R, then the perpendicular vector from the line L to the point r is given by

$$\Delta \mathrm{r}_i^{\perp} = (\mathrm{r}_i - \mathrm{R}) - (\mathrm{S}\cdot(\mathrm{r}_i - \mathrm{R}))\mathrm{S} = [[I] - [\mathrm{SS}^T]](\Delta \mathrm{r}_i),$$

where [I] is the identity matrix and [S S^T] is the outer product matrix formed from the unit vector S along the line L. Recall that skew-symmetric matrix [S] is constructed so that [S]y=S x y. The matrix [I-SS^T] in this equation subtracts the component of Δr=r-R that is parallel to S.

The previous sections show that in computing the moment of inertia matrix this operator yields a similar operator using the components of the vector Δr that is

$$[I\,|\Delta \mathrm{r}|^2 - \Delta \mathrm{r} \Delta \mathrm{r}^T].$$

It is helpful to keep the following identities in mind In order to compare the equations that define the inertia tensor and the inertia matrix.

Let [R] be the skew symmetric matrix associated with the position vector R=(x, y, z), then the product in the inertia matrix becomes

$$-[R]^2 = -\begin{bmatrix} 0 & -z & y \\ z & 0 & -x \\ -y & x & 0 \end{bmatrix}^2 = \begin{bmatrix} y^2+z^2 & -xy & -xz \\ -yx & x^2+z^2 & -yz \\ -zx & -zy & x^2+y^2 \end{bmatrix}.$$

This can be viewed as another way of computing the perpendicular distance from an axis to a point, because the matrix formed by the outer product [R R^T] yields the identify

$$-[R]^2 = |\mathrm{R}|^2\,[I] - [\mathrm{RR}^T] = \begin{bmatrix} x^2+y^2+z^2 & 0 & 0 \\ 0 & x^2+y^2+z^2 & 0 \\ 0 & 0 & x^2+y^2+z^2 \end{bmatrix} - \begin{bmatrix} x^2 & xy & xz \\ yx & y^2 & yz \\ zx & zy & z^2 \end{bmatrix},$$

where [I] is the 3x3 identity matrix.

Also notice, that

$$|\mathrm{R}|^2 = \mathrm{R}\cdot\mathrm{R} = \mathrm{tr}[\mathrm{RR}^T],$$

where *tr* denotes the sum of the diagonal elements of the outer product matrix, known as its trace.

The Inertia Matrix in Different Reference Frames

The use of the inertia matrix in Newton's second law assumes its components are computed relative to axes parallel to the inertial frame and not relative to a body-fixed reference frame. This means that as the body moves the components of the inertia matrix change with time. In contrast, the components of the inertia matrix measured in a body-fixed frame are constant.

Body Frame Inertia Matrix

Let the body frame inertia matrix relative to the centre of mass be denoted $[I_C^B]$, and define the orientation of the body frame relative to the inertial frame by the rotation matrix [A], such that,

$$\mathrm{x} = [A]\mathrm{y},$$

where vectors y in the body fixed coordinate frame have coordinates x in the inertial frame. Then, the inertia matrix of the body measured in the inertial frame is given by

$$[I_C] = [A][I_C^B][A^T].$$

Notice that [A] changes as the body moves, while $[I_C^B]$ remains constant.

Principal Axes

Measured in the body frame the inertia matrix is a constant real symmetric matrix. A real symmetric matrix has the eigendecomposition into the product of a rotation matrix [Q] and a diagonal matrix $[\Lambda]$, given by

$$[I_C^B] = [Q][\Lambda][Q^T],$$

where

$$[\Lambda] = \begin{bmatrix} I_1 & 0 & 0 \\ 0 & I_2 & 0 \\ 0 & 0 & I_3 \end{bmatrix}.$$

The columns of the rotation matrix [Q] define the directions of the principal axes of the body, and the constants I_1, I_2 and I_3 are called the principal moments of inertia. This result was first shown by J. J. Sylvester (1852), and is a form of Sylvester's law of inertia.

For bodies with constant density an axis of rotational symmetry is a principal axis.

Inertia Ellipsoid

The moment of inertia matrix in body-frame coordinates is a quadratic form that defines a surface in the body called Poinsot's ellipsoid. Let $[\Lambda]$ be the inertia matrix relative to the centre of mass aligned with the principal axes, then the surface

$$\mathrm{x}^T[\Lambda]\mathrm{x} = 1,$$

or

$$I_1x^2 + I_2y^2 + I_3z^2 = 1,$$

defines an ellipsoid in the body frame. Write this equation in the form,

$$\frac{x^2}{(1/\sqrt{I_1})^2}+\frac{y^2}{(1/\sqrt{I_2})^2}+\frac{z^2}{(1/\sqrt{I_3})^2}=1,$$

to see that the semi-principal diameters of this ellipsoid are given by

$$a=\frac{1}{\sqrt{I_1}},\quad b=\frac{1}{\sqrt{I_2}},\quad c=\frac{1}{\sqrt{I_3}}.$$

Let a point x on this ellipsoid be defined in terms of its magnitude and direction, x= | x | n, where n is a unit vector. Then the relationship presented above, between the inertia matrix and the scalar moment of inertia I_n around an axis in the direction n, yields

$$\mathrm{x}^T[\Lambda]\mathrm{x}=|\mathrm{x}|^2\,\mathrm{n}^T[\Lambda]\mathrm{n}=|\mathrm{x}|^2\,I_n=1.$$

Thus, the magnitude of a point x in the direction n on the inertia ellipsoid is

$$|\mathrm{x}|=\frac{1}{\sqrt{I_n}}.$$

List of Moments of Inertia

The mass moment of inertia, usually denoted *I*, measures the extent to which an object resists rotational acceleration about an axis, and is the rotational analogue to mass. Mass moments of inertia have units of dimension mass × length^2. It should not be confused with the second moment of area, which is used in bending calculations.

Geometrically simple objects have moments of inertia that can be expressed mathematically, but it may not be straightforward to symbolically express the moment of inertia of more complex bodies.

The following moments of inertia assume constant density throughout the object, and the axis of rotation is taken to be through the centre of mass, unless otherwise specified.

Theorems of Moment of Inertia

In physics, the parallel axis theorem, also known as Huygens–Steiner theorem after Christiaan Huygens and Jakob Steiner, can be used to determine the mass moment of inertia or the second moment of area of a rigid body about any axis, given the body's moment of inertia about a parallel axis through the object's centre of mass and the perpendicular distance between the axes.

Mass Moment of Inertia

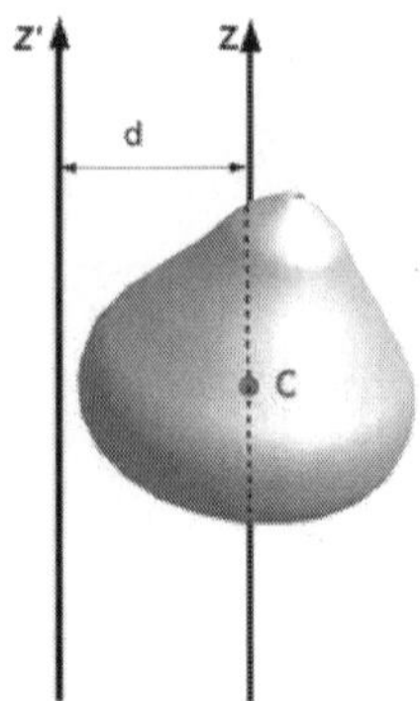

Figure: *The mass moment of inertia of a body around an axis can be determined from the mass moment of inertia around a parallel axis through the centre of mass.*

Suppose a body of mass m is made to rotate about an axis z passing through the body's centre of mass. The body has a moment of inertia I_{cm} with respect to this axis.

The parallel axis theorem states that if the body is made to rotate instead about a new axis $z2$ which is parallel to the first axis and displaced from it by a distance r, then the moment of inertia I with respect to the new axis is related to I_{cm} by

$$I = I_{cm} + mr^2.$$

Explicitly, r is the perpendicular distance between the axes z and $z2$.

The parallel axis theorem can be applied with the stretch rule and perpendicular axis theorem to find moments of inertia for a variety of shapes.

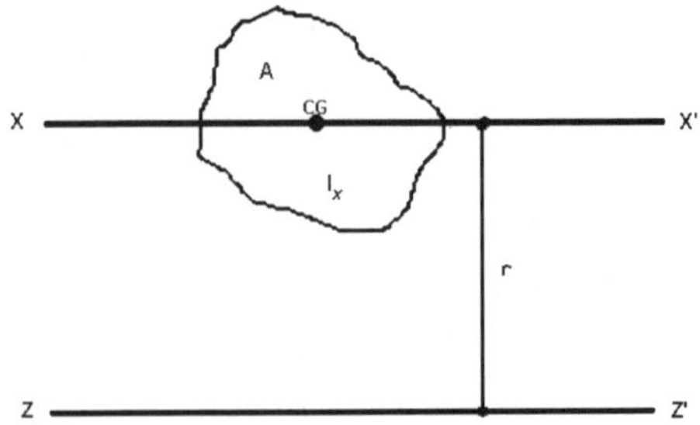

Figure: *Parallel axes rule for area moment of inertia*

Proof

We may assume, without loss of generality, that in a Cartesian coordinate system the perpendicular distance between the axes lies along the x-axis and that the centre of mass lies at the origin. The moment of inertia relative to the z-axis is

$$I_{cm} = \int (x^2 + y^2) dm.$$

The moment of inertia relative to the axis $z2$, which is a perpendicular distance r along the x-axis from the centre of mass, is

$$I = \int \left[(x-r)^2 + y^2 \right] dm$$

Expanding the brackets yields

$$I = \int (x^2 + y^2) dm + r^2 \int dm - 2r \int x dm.$$

The first term is I_{cm}, the second term becomes mr^2, and the final term is zero since the origin of the coordinates is at the centre of mass. So, the equation becomes:

$$I = I_{cm} + mr^2.$$

Tensor Generalization

The parallel axis theorem can be generalized to calculations involving the inertia tensor. Let I_{ij} denote the inertia tensor of a body as calculated at the centre of mass. Then the inertia tensor J_{ij} as calculated relative to a new point is

$$J_{ij} = I_{ij} + m\left(|\mathrm{R}|^2 \delta_{ij} - R_i R_j \right),$$

where $\mathrm{R} = R_1 \hat{\mathrm{x}} + R_2 \hat{\mathrm{y}} + R_3 \hat{\mathrm{z}}$ is the displacement vector from the centre of mass to the new point, and d_{ij} is the Kronecker delta.

For diagonal elements (when $i = j$), displacements perpendicular to the axis of rotation results in the above simplified version of the parallel axis theorem.

The generalized version of the parallel axis theorem can be expressed in coordinate-free notation as

$$\mathrm{J} = \mathrm{I} + m\left[(\mathrm{R} \cdot \mathrm{R}) \mathrm{E}_3 - \mathrm{R} \otimes \mathrm{R} \right],$$

where E_3 is the 3 × 3 identity matrix and $\otimes$ is the outer product.

Area Moment of Inertia

The parallel axes rule also applies to the second moment of area (area moment of inertia) for a plane region D:

$$I_z = I_x + Ar^2,$$

where I_z is the area moment of inertia of D relative to the parallel axis, I_x is the area moment of inertia of D relative to its centroid, A is the area of the plane region D, and r is the distance from the new axis z to the centroid of the plane region D. The centroid of D coincides with the centre of gravity of a physical plate with the same shape that has uniform density.

Polar Moment of Inertia for Planar Dynamics

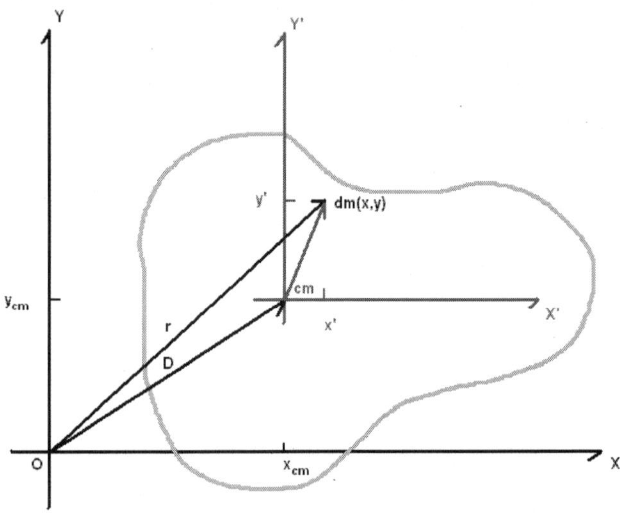

Figure: *Polar moment of inertia of a body around a point can be determined from its polar moment of inertia around the centre of mass.*

The mass properties of a rigid body that is constrained to move parallel to a plane are defined by its centre of mass R = (x, y) in this plane, and its polar moment of inertia I_R around an axis through R that is perpendicular to the plane. The parallel axis theorem provides a convenient relationship between the moment of inertia I_S around an arbitrary point S and the moment of inertia I_R about the centre of mass R. Recall that the centre of mass R has the property

$$\int_V \rho(\mathrm{r})(\mathrm{r} - \mathrm{R})dV = 0,$$

where r is integrated over the volume V of the body. The polar moment of inertia of a body undergoing planar movement can be computed relative to any reference point S,

$$I_S = \int_V \rho(\mathrm{r})(\mathrm{r} - \mathrm{S})\cdot(\mathrm{r} - \mathrm{S})dV,$$

where S is constant and r is integrated over the volume V.

In order to obtain the moment of inertia I_S in terms of the moment of inertia I_R, introduce the vector d from S to the centre of mass R,

$$I_S = \int_V \rho(\mathrm{r})(\mathrm{r}-\mathrm{R}+\mathrm{d})\cdot(\mathrm{r}-\mathrm{R}+\mathrm{d})dV$$

$$= \int_V \rho(\mathrm{r})(\mathrm{r}-\mathrm{R})\cdot(\mathrm{r}-\mathrm{R})dV + 2\mathrm{d}\cdot\left(\int_V \rho(\mathrm{r})(\mathrm{r}-\mathrm{R})dV\right) + \left(\int_V \rho(\mathrm{r})dV\right)\mathrm{d}\cdot\mathrm{d}.$$

The first term is the moment of inertia I_R, the second term is zero by definition of the centre of mass, and the last term is the total mass of the body times the square magnitude of the vector d. Thus,

$$I_S = I_R + Md^2,$$

which is known as the parallel axis theorem.

Moment of Inertia Matrix

The inertia matrix of a rigid system of particles depends on the choice of the reference point. There is a useful relationship between the inertia matrix relative to the centre of mass R and the inertia matrix relative to another point S. This relationship is called the parallel axis theorem.

Consider the inertia matrix $[I_S]$ obtained for a rigid system of particles measured relative to a reference point S, given by

$$[I_S] = -\sum_{i=1}^{n} m_i[r_i - S][r_i - S],$$

where r_i defines the position of particle P_i, $i = 1, \ldots, n$. Recall that $[r_i - S]$ is the skew-symmetric matrix that performs the cross product,

$$[r_i - S]\mathrm{y} = (\mathrm{r}_i - \mathrm{S}) \times \mathrm{y},$$

for an arbitrary vector y.

Let R be the centre of mass of the rigid system, then

$$\mathrm{R} = (\mathrm{R}-\mathrm{S}) + \mathrm{S} = \mathrm{d} + \mathrm{S},$$

where d is the vector from the reference point S to the centre of mass *R*. Use this equation to compute the inertia matrix,

$$[I_S] = -\sum_{i=1}^{n} m_i[r_i - R + d][r_i - R + d].$$

Expand this equation to obtain

$$[I_S] = \left(-\sum_{i=1}^{n} m_i[r_i - R][r_i - R]\right) + \left(-\sum_{i=1}^{n} m_i[r_i - R]\right)[d] + [d]\left(-\sum_{i=1}^{n} m_i[r_i - R]\right) + \left(-\sum_{i=1}^{n} m_i\right)[d][d].$$

The first term is the inertia matrix $[I_R]$ relative to the centre of mass. The second and third terms are zero by definition of the centre of mass *R*,

$$\sum_{i=1}^{n} m_i(\mathrm{r}_i - \mathrm{R}) = 0.$$

And the last term is the total mass of the system multiplied by the square of the skew-symmetric matrix [*d*] constructed from d.

The result is the parallel axis theorem,

$$[I_S] = [I_R] - M[d]^2,$$

where *d* is the vector from the reference point S to the centre of mass *R*.

7

Capstan Equation

The capstan equation or belt friction equation, also known as Eytelwein's formula, relates the hold-force to the load-force if a flexible line is wound around a cylinder (a bollard, a winch or a capstan) . Because of the interaction of frictional forces and tension, the tension on a line wrapped around a capstan may be different on either side of the capstan. A small *holding* force exerted on one side can carry a much larger *loading* force on the other side; this is the principle by which a capstan-type device operates. For instance in rock climbing with so-called top-roping, a lighter person can hold (belay) a heavier person due to this effect. The formula is:

$$T_{\text{load}} = T_{\text{hold}}\, e^{\mu\phi}$$

where T_{load} is the applied tension on the line, T_{hold} is the resulting force exerted at the other side of the capstan, μ is the coefficient of friction between the rope and capstan materials, and ϕ is the total angle swept by all turns of the rope, measured in radians (i.e., with one full turn the angle $\phi = 2\pi$).

Several assumptions must be true for the formula to be valid:

1. The rope is on the verge of full sliding, i.e. T_{load} is the maximum load that one can hold. Smaller loads can be held as well, resulting in a smaller *effective* contact angle ϕ.
2. It is important that the line is not rigid, in which case significant force would be lost in the bending of the line tightly around the cylinder. (The equation must be modified for this case.) For instance a Bowden cable is to some extent rigid and doesn't obey the principles of the Capstan equation.
3. The line is non-elastic.

It can be observed that the force gain grows exponentially with the coefficient of friction, the number of turns around the cylinder, and the angle of contact. Note that the radius of the cylinder has no influence on the force gain. The table $e^{\mu\phi}$ below lists values of the factor based on the number of turns and coefficient of friction μ.

Number of turns	Coefficient of friction μ 0.1	0.2	0.3	0.4	0.5	0.6	0.7
1	1.9	3.5	6.6	12	23	43	81
2	3.5	12	43	152	535	1881	6661
3	6.6	43	286	1881	12392	81612	437503
4	12	152	1881	23228	286751	3540026	43702631
5	23	535	12392	286751	6635624	153552935	3553321281

From the table it is evident why one seldom sees a sheet (a rope to the loose side of a sail) wound more than three turns around a winch. The force gain would be extreme besides being counter-productive since there is risk of a riding turn, result being that the sheet will foul, form a knot and not run out when eased (by slacking grip on the tail (free end), or in land talk, one lets go of the hold end. It is both ancient and modern practice for anchor capstans and jib winches to be slightly flared out at the base, rather than cylindrical, to prevent the rope (anchor warp or sail sheet) from sliding down. The rope wound several times around the winch can slip upwards gradually, with little risk of a riding turn, provided it is tailed (loose end is pulled clear), by hand or a self-tailer.

For instance, the factor 153552935 means, in theory, that a newborn baby would be capable of holding the weight of two USS *Nimitz* supercarriers (97 000 ton each, but for the baby it would be only a little more than 1 kg).

Proof of the Capstan Equation

1. Circular coordinates

$$\varphi = \frac{s}{r} \qquad d\varphi = \frac{ds}{r} \qquad \frac{d\varphi}{ds} = \frac{1}{r} \qquad (1), (2), (3)$$

Let s_u and n_u denote unit vectors;

$$s_u = -\sin\varphi \cdot x_u + \cos\varphi \cdot y_u \qquad (4)$$

$$n_u = \cos\varphi \cdot x_u + \sin\varphi \cdot y_u \qquad (5)$$

Then from (5)

$$\frac{d}{d\varphi} s_u = -\cos\varphi \cdot x_u - \sin\varphi \cdot y_u = -n_u \qquad (6)$$

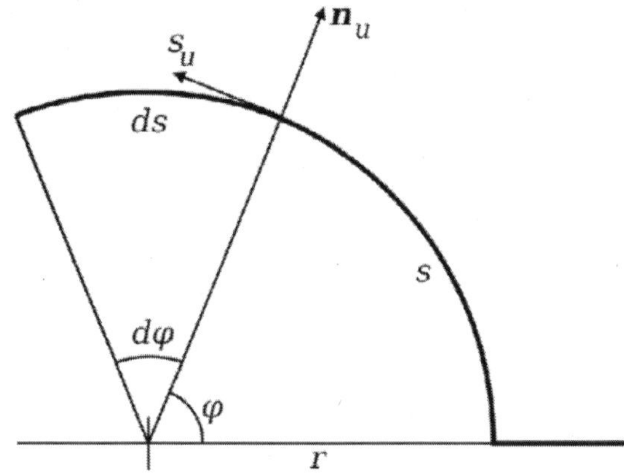

$$\frac{d}{ds}=\frac{d}{d\varphi}\cdot\frac{d\varphi}{ds}=\frac{d}{d\varphi}\cdot\frac{1}{r}=\frac{1}{r}\cdot\frac{d}{d\varphi} \tag{7}$$

From (6) and (7), it follows that

$$\frac{d}{ds}\mathrm{s}_u=\frac{1}{r}\cdot\frac{d}{d\varphi}\mathrm{s}_u=-\frac{1}{r}\cdot\mathrm{n}_u. \tag{8}$$

2. *Forces on Cordage in General:* Now, let's study a piece of cord in general, subject to an arbitrary force. Let s denote the length of the cord and let the force *per unit length* be $\mathrm{q}(s)$. Consider a short piece Δs of the cord and introduce the cross-sectional force $\mathrm{T}(s)$. Balancing the forces, we get

$$\mathrm{T}(s+\Delta s)-\mathrm{T}(s)+\mathrm{q}(s)\cdot\Delta s=0, (9)$$

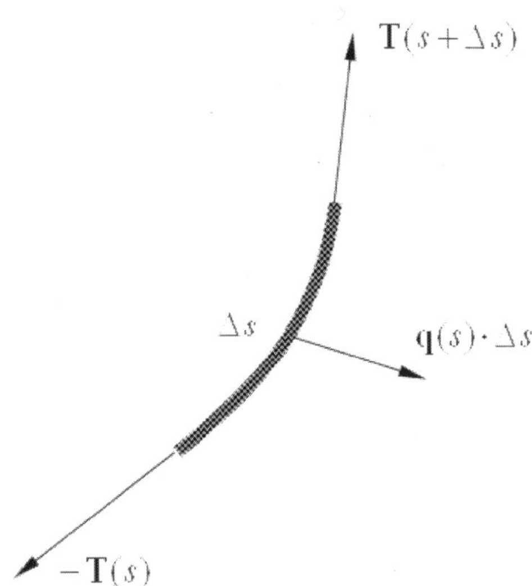

$$\frac{\mathrm{T}(s+\Delta s)-\mathrm{T}(s)}{\Delta s}=-\mathrm{q}(s). \tag{10}$$

Letting $\Delta s \to 0$, we conclude that

$$\frac{d}{ds}\mathrm{T}(s)=-\mathrm{q}(s). \tag{11}$$

3. *A Line Around a Capstan:* A line is wound around a cylinder(a bollard or a capstan). In this case the curvature of the line is circular which makes the problem easier. Let s be the length Δs of the line from a point A where the line makes contact with the cylinder. At the point on the short piece of the line acts a force from the cylinder that can be subdivided into a *tangential* component $t\Delta s$ (friction) and a *normal* component $n\Delta s$. That is to say that

$$\mathrm{q}(s)=t\cdot\mathrm{s}_u+n\cdot\mathrm{n}_u \tag{12}$$

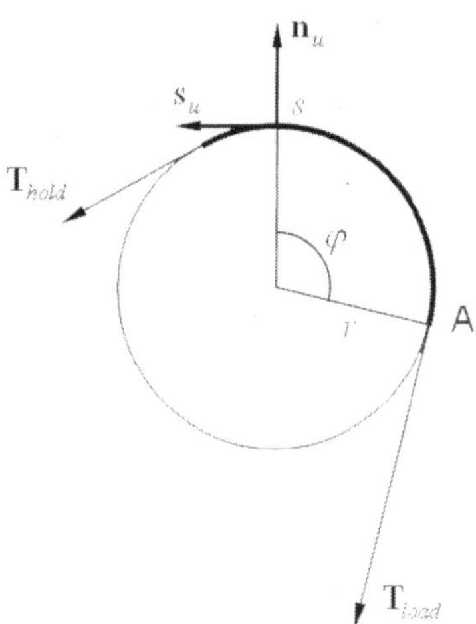

With the cross-sectional force $\mathrm{T}(s)$ (which is tangential) we get

$$\mathrm{T}(s)=T(s)\cdot\mathrm{s}_u \tag{13}$$

From (11), (12) and (13), it follows that

$$\frac{d}{ds}T(s)\cdot\mathrm{s}_u=-t\cdot\mathrm{s}_u-n\cdot\mathrm{n}_u \tag{14}$$

Derivative of a product and (8) imply that

$$\frac{d}{ds}T(s)\cdot\mathrm{s}_u=\frac{dT(s)}{ds}\cdot\mathrm{s}_u+T(s)\cdot\frac{d}{ds}\mathrm{s}_u$$

$$=\frac{dT(s)}{ds}\cdot\mathrm{s}_u-\frac{T(s)}{r}\cdot\mathrm{n}_u=-t\cdot\mathrm{s}_u-n\cdot\mathrm{n}_u \tag{15}$$

Identifying components in (15), we get

$$\frac{dT(s)}{ds} = -t \tag{16}$$

and

$$\frac{T(s)}{r} = n. \tag{17}$$

Dividing (16) by (17), we get

$$\frac{dT(s)}{ds} / \frac{T(s)}{r} = -\frac{t}{n} \tag{18}$$

From (18) and reciprocal of (2), we get

$$\text{LHS} = \frac{dT(s)}{T(s)} \cdot \frac{r}{ds} = \frac{dT(s)}{T(s)} \cdot \frac{1}{d\varphi} = \frac{1}{T(S)} \cdot \frac{dT(s)}{d\varphi}. \tag{19}$$

From (18) and (19) it follows that

$$\frac{1}{T(s)} \cdot \frac{dT(s)}{d\varphi} = -\frac{t}{n}. \tag{20}$$

Let $\mu = \frac{t}{n}$ (21) be the *coefficient of friction* (no slip). Then

$$\frac{1}{T} \cdot \frac{dT}{d\varphi} = -\mu \qquad (22) : \Rightarrow$$

$$\frac{1}{T} \cdot dT = -\mu \cdot d\varphi \tag{23}$$

Integration of (23) yields

$$\int_{T_{\text{load}}}^{T_{\text{hold}}} \frac{1}{T} \cdot dT = \int_0^{\phi} -\mu \cdot d\varphi \tag{24}$$

$$\ln T_{\text{hold}} - \ln T_{\text{load}} = \ln \frac{T_{\text{hold}}}{T_{\text{load}}} = -\mu \cdot \phi \tag{25}$$

$$\frac{T_{\text{hold}}}{T_{\text{load}}} = e^{-\mu \cdot \phi} \tag{26}$$

Finally,

$$T_{\text{hold}} = T_{\text{load}} \cdot e^{-\mu \cdot \phi} \quad \text{or} \quad T_{\text{load}} = T_{\text{hold}} \cdot e^{\mu \cdot \phi}$$

Friction Coefficient

There are certain factors that help determine the value of the friction coefficient. These determining factors are:

- Belting material used – The age of the material also plays a part, where worn out and older material tends to be more rough and therefore experience greater friction when sliding.
- Construction of the drive-pulley system – This involves strength and stability of the material used, like the pulley, and how greatly it will oppose the motion of the belt or rope.
- Conditions in which the test is being – The friction between the belt and pulley will decrease immensely if the belt happens to be muddy or wet, as it acts as lubricant to the force. This also applies to extremely dry or warm conditions which will evaporate any water naturally found in the belt, making friction much greater.
- Overall design of the setup – The setup involves the initial conditions of the construction, such as the angle which the belt is wrapped around and the maximum amount of tension that can be sustained by the belt.

Applications

An understanding of belt friction is essential for sailing crews and mountain climbers. Their professions require being able to maximize the amount of weight a rope with a certain tension capacity can hold versus the amount of wraps around a pulley. Too many revolutions around a pulley make it inefficient to retract or release rope, and too few may cause the rope to slip. Misjudging the ability of a rope to sustain itself against a certain force will ultimately lead to failure or serious injury.

Laws of Friction

Laws of Dry Friction and Coefficients of Friction[* Internal error: Invalid file format. | In-line.WMF *]

Equations

(Eq1) $F_m = \mu_s N$ Magnitude of the maximum static friction force

(Eq2) $F_k = \mu_k N$ Magnitude of the kinetic friction force

Nomenclature

symbol	*description*
F_m	maximum force
F_k	kinetic friction force
μ_s	coefficient of static friction
μ_k	coefficient of kinetic friction
N	normal component of the reaction surface

Explanation

Friction is both good and bad: it allows objects to move, such as in walking or driving, yet can result in a lot of wasted energy through the generation of heat, such as in engines, and in other areas. Friction can be understood simply by observing the following picture of a cinder block sitting on asphalt:

Zooming in, it can be seen that there are irregularities in the surfaces of both materials, and when they are moved against each other, these small bits are pushing against each other. It is the summation of all of these horizontal forces causing resistance that make it hard to push an object with a lot of friction. When an object is pushed the bits may break off, or they may bump over each other. The weight of the object, or any vertical forces that are pushing down on the object, will influence the friction because as vertical downward force is increased, it becomes harder to overcome those horizontal forces between all the little small bits interconnected with each other between the contacting surfaces.

The laws of dry friction may be explained with a wooden block of weight W that is placed on a horizontal concrete plane surface.

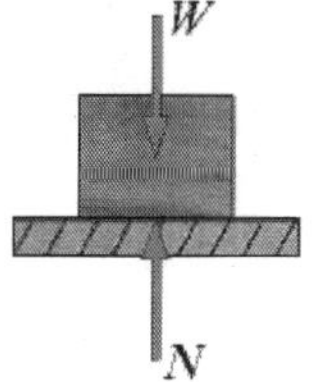

The forces acting on the block are its weight W and the reaction of the surface. Since the weight has no horizontal component, the reaction of the surface also has no horizontal component. The reaction is therefore normal to the surface and is represented by N.

Now a horizontal force P is applied to the block as shown in the following Figure:

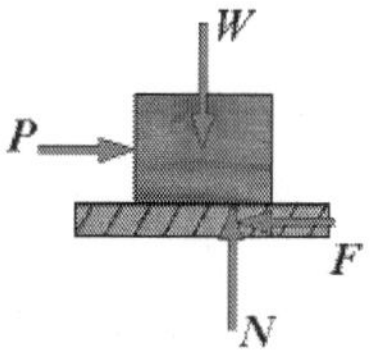

If P is small enough, the block will not move. This only means that some other horizontal force must therefore exist, which balances P. This other force is the static-friction force F, which is actually the resultant of numerous forces acting over the entire surface of contact between the block and the plane, as indicated in the first picture in this lesson. It is generally assumed that these forces are due to the irregularities of the surfaces in contact and, to a certain extent, to molecular attraction.

If the force P is increased, the friction force F also increases, continuing to oppose P until its magnitude reaches a certain maximum value F_m, as shown in the following plot:

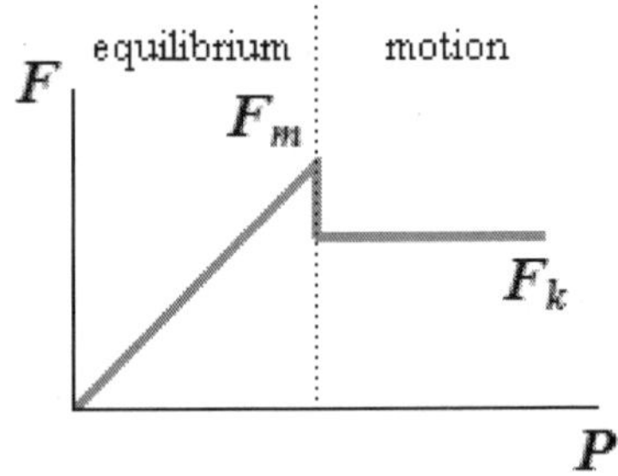

If P is further increased, the friction force cannot balance it any more and the block starts sliding. A more accurate representation of the previous plot may be the following:

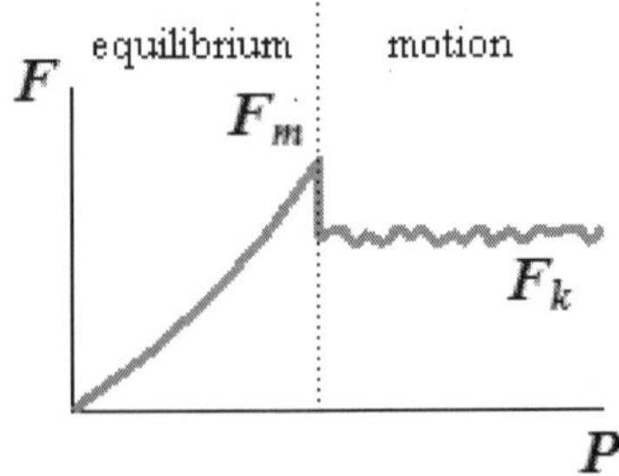

F and *P* are relatively proportional until F_m is reached, and motion begins. F_k may be shown as a wavy and bumpy line because the frictional force varies at the micro-level interactions, as seen in the first picture in this lesson, which may not be consistent.

Notice in the following figure that, as the magnitude *F* of the friction force increases from 0 to F_m, the point of application *a* of the resultant *N* of the normal forces of contact moves to the right, so that the couples formed, respectively, by P and F and by W and N remain balanced.

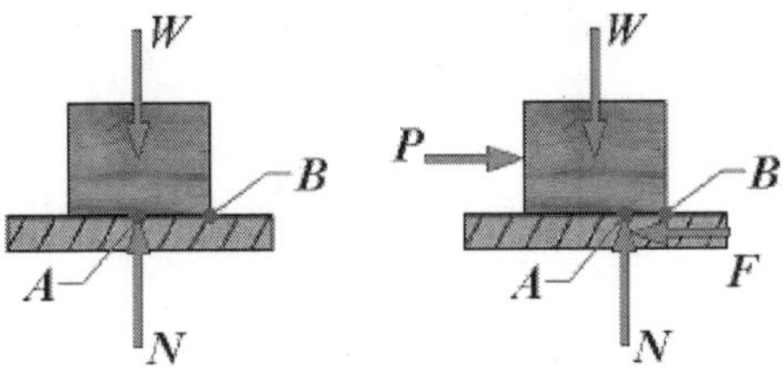

If N reaches point *b*, which is fixed at the bottom corner of the block, before F reaches its maximum value F_m, the couples will no longer be balanced and the block will tip about point *b* before the block can start sliding.

As soon as the block has been set in motion, the magnitude of F drops from F_m to a lower value F_k as shown below and in the above plots:

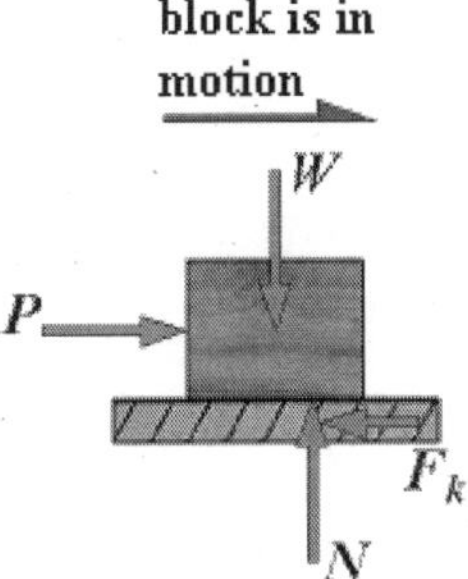

This is because there is less interpenetration between the irregularities of the surfaces in contact when these surfaces move with respect to each other. From then on, the block keeps sliding with increasing velocity while the friction force is now denoted by F_k, which is called the kinetic friction force. This can be observed in the plot above. The kinetic friction force remains approximately constant throughout the motion of the block.

Experimental evidence shows that the maximum value F_m of the static friction force is proportional to the normal component N of the reaction of the surface. The equation is:

(Eq1) $F_m = \mu_s N$

Where μ_s is a constant called the coefficient of static friction. Similarly, the magnitude F_k of the kinetic friction force may be put in the form:

(Eq2) $F_k = \mu_k N$

Where m_k is a constant called the coefficient of kinetic friction. The coefficients of friction m_s and m_k do not depend upon the area of the surfaces in contact. Both coefficients, however, depend strongly on the nature of the surfaces in contact. Since they also depend upon the exact condition of the surfaces, their value is seldom known with an accuracy greater than 5 percent. The corresponding values of the coefficient of kinetic friction would be about 25 percent smaller.

It is presumed that four different situations can occur when a rigid body is in contact with a horizontal surface:

1.	No friction ($P_x = 0$)	$F = 0$ $N = P + W$	The forces applied to the body do not tend to move it along the surface of contact; there is not friction force.
2.	No motion ($P_x < F_m$)	$F = P_x$ $F < \mu_s N$ $N = P_y + W$	The applied forces tend to move the body along the surface of contact but are not large enough to set it in motion. The friction force F which has developed can be found by solving the equations of equilibrium for the body. Since there is no evidence that F has reached its maximum value, the equation $F_m = \mu_s N$ cannot be used to determine the friction force.
3.	Motion impending ($P_x = F_m$)	$F_m = P_x$ $F_m = \mu_s N$ $N = P_y + W$	The applied forces are such that the body is just about to slide, that is, motion is impending. The friction force F has reached its maximum value F_m and, together with the normal force N, balances the applied forces. Both the equations of equilibrium and the equation $F_m = \mu_s N$ can be used. Also note that the friction force has a sense opposite to the sense of impending motion.
4.	Motion ($P_x > F_m$)	$F_k < P_x$ $F_k = \mu_k N$ $N = P_y + W$	The body is sliding under the action of the applied forces, and the equations of equilibrium do not apply any more. However, F is now equal to F_k and the equation $F_k = \mu_k N$ may be used. The sense of F_k is opposite to the sense of motion.

Motion of Bodies

In physics, motion is a change in position of an object with respect to time and its reference point. Motion is typically described in terms of displacement, direction, velocity, acceleration, and time. Motion is observed by attaching a frame of reference to a body and measuring its change in position relative to that frame.

A body which does not move is said to be *at rest*, *motionless*, *immobile*, *stationary*, or to have constant (time-invariant) position. An object's motion cannot change unless it is acted upon by a force, as described by Newton's first law. An object's momentum is directly related to the object's mass and velocity, and the total momentum of all objects in a closed system (one not affected by external forces) does not change with time, as described by the law of conservation of momentum.

As there is no absolute frame of reference, *absolute motion* cannot be determined. Thus, everything in the universe can be considered to be moving.

More generally, the term motion signifies a continuous change in the configuration of a physical system. For example, one can talk about motion of a wave or a quantum particle (or any other field) where the configuration consists of probabilities of occupying specific positions.

Laws of Motion

In physics, motion in the universe is described through two sets of apparently contradictory laws of mechanics. Motions of all large scale and familiar objects in the universe (such as projectiles, planets, cells, and humans) are described by classical mechanics. Whereas the motion of very small atomic and sub-atomic objects is described by quantum mechanics.

Classical Mechanics

Classical mechanics is used for describing the motion of macroscopic objects, from projectiles to parts of machinery, as well as astronomical objects, such as spacecraft, planets, stars, and galaxies. It produces very accurate results within these domains, and is one of the oldest and largest subjects in science, engineering, and technology.

Classical mechanics is fundamentally based on Newton's Laws of Motion. These laws describe the relationship between the forces acting on a body and the motion of that body. They were first compiled by Sir Isaac Newton in his work *Philosophiæ Naturalis Principia Mathematica*, first published on July 5, 1687. His three laws are:

1. In the absence of a net external force, a body either is at rest or moves with constant velocity.
2. The net external force on a body is equal to the mass of that body times its acceleration; F = *m*a. Alternatively, the acceleration is directly proportional to the force causing it, and inversely proportional to the mass.
3. Whenever one body exerts a force F onto a second body, the second body exerts the force –F on the first body. F and –F are equal in magnitude and opposite in sense.

Newton's three laws of motion, along with his law of universal gravitation, explain Kepler's laws of planetary motion, which were the first to accurately provide a mathematical model for understanding orbiting bodies in outer space. This explanation unified the motion of celestial bodies and motion of objects on earth.

Classical mechanics was later further enhanced by Albert Einstein's special relativity and general relativity. Special relativity explains the motion of objects with a high velocity, approaching the speed of light; general relativity is employed to handle gravitational motion at a deeper level.

Quantum Mechanics

Quantum mechanics is a set of principles describing physical reality at the atomic level of matter (molecules and atoms) and the subatomic (electrons, protons, and even smaller particles). These descriptions include the simultaneous wave-like and particle-like behaviour of both matter and radiation energy, this is described in the wave–particle duality.

In contrast to classical mechanics, where accurate measurements and predictions can be calculated about location and velocity, in the quantum mechanics of a subatomic particle, one can never specify its state, such as its simultaneous location and velocity, with complete certainty (this is called the Heisenberg uncertainty principle).

In addition to describing the motion of atomic level phenomena, quantum mechanics is useful in understanding some large scale phenomenon such as superfluidity, superconductivity, and biological systems, including the function of smell receptors and the structures of proteins.

Kinematics

Kinematics applies geometry to the analysis of movement, or motion, of a mechanical system. Rest and motion are very relative terms as far as the frames of reference are concerned. So, while proceeding with any of the problems in kinematics one should be very careful about the frame of reference. The rotation and sliding movement central in a mechanical system is modelled mathematically as Euclidean, or rigid, transformations. The set of rigid transformations in three dimensional space forms a Lie group, denoted as SE(3).

Planar Motion

While all motion in a mechanical system occurs in three dimensional space, planar motion can be analyzed using plane geometry, if all point trajectories are parallel to a plane. In this case the system is called a *planar mechanism* (or robot). The kinematic analysis of planar mechanisms uses the subset of SE(3) consisting of planar rotations and translations, denoted SE(2).

The group SE(2) is three dimensional, which means that every position of a body in the plane is defined by three parameters. The parameters are often the x and y coordinates of the origin of a coordinate frame in M measured from the origin of a coordinate frame in F, and the angle measured from the x-axis in F to the x-axis in M. This is described saying a body in the plane has three degrees-of-freedom. SE(2) is the configuration space for a planar body, and a planar motion is a curve in this space.

Spherical Motion

It is possible to construct a mechanical system such that the point trajectories in all components lie in concentric spherical shells around a fixed point. An example is the gimbaled gyroscope. These devices are called *spherical mechanisms*. Spherical mechanisms are constructed by connecting links with hinged joints such that the axes of each hinge passes through the same point. This point becomes centre of the concentric spherical shells. The movement of these mechanisms is characterized by the group SO(3) of rotations in three dimensional space. Other examples of spherical mechanisms are the automotive differential and the robotic wrist. Select this link for an animation of a Spherical deployable mechanism.

The rotation group SO(3) is three dimensional. An example of the three parameters that specify a spatial rotation are the roll, pitch and yaw angles used to define the orientation of an aircraft. SO(3) is the configuration space for a rotating body, and a spherical motion is a curve in this space.

Spatial Motion

A mechanical system in which a body moves through a general spatial movement is called a *spatial mechanism*. An example is the RSSR linkage, which can be viewed as a four-bar linkage in which the hinged joints of the coupler link are replaced by rod ends, also called spherical joints or ball joints. The rod ends allow the input and output cranks of the RSSR linkage to be misaligned to the point that they lie in different planes, which causes the coupler link to move in a general spatial movement. Robot arms, Stewart platforms, and humanoid robotic systems are also examples of spatial mechanisms.

Select this link for an animation of Bennett's linkage, which is a spatial mechanism constructed from four hinged joints.

The group SE(3) is six dimensional, which means the position of a body in space is defined by six parameters. Three of the parameters define the origin of the moving reference frame relative to the fixed frame. Three other parameters define the orientation of the moving frame relative to the fixed frame. SE(3) is the configuration space for a body moving in space, and a spatial motion is a curve in this space.

Applications

Friction is an important factor in many engineering disciplines.

Transportation

- Rail adhesion refers to the grip wheels of a train have on the rails.

- Road slipperiness is an important design and safety factor for automobiles
  - o Split friction is a particularly dangerous condition arising due to varying friction on either side of a car.
  - o Road texture affects the interaction of tires and the driving surface.

Measurement

- A tribometer is an instrument that measures friction on a surface.
- A profilograph is a device used to measure pavement surface roughness.

Household Usage

- Friction is used to ignite matchstick (Friction between the head of a matchstick and the rubbing surface of the match box).

Instances where Friction is Advantageous

- Friction between meteorite and air helps to increase its temperature and ignite it, protecting the living things on Earth.
- Finger prints and palm prints on hand help to hold up objects easily in human and ape hands.
- Foot prints increase friction and helps to stand up.
- Automobile brakes inherently rely on friction, slowing a vehicle by converting its kinetic energy into heat. Incidentally, dispersing this large amount of heat safely is one technical challenge in designing brake systems.

Reducing Friction

Devices

Devices such as wheels, ball bearings, roller bearings, and air cushion or other types of fluid bearings can change sliding friction into a much smaller type of rolling friction. Many thermoplastic materials such as nylon, HDPE and PTFE are commonly used in low friction bearings. They are especially useful because the coefficient of friction falls with increasing imposed load. For improved wear resistance, very high molecular weight grades are usually specified for heavy duty or critical bearings.

Lubricants

A common way to reduce friction is by using a lubricant, such as oil, water, or grease, which is placed between the two surfaces, often dramatically lessening the coefficient of friction. The science of friction and lubrication is called tribology. Lubricant technology is when lubricants are mixed with the application of science, especially to industrial or commercial objectives.

Superlubricity, a recently-discovered effect, has been observed in graphite: it is the substantial decrease of friction between two sliding objects, approaching zero levels. A very small amount of frictional energy would still be dissipated.

Lubricants to overcome friction need not always be thin, turbulent fluids or powdery solids such as graphite and talc; acoustic lubrication actually uses sound as a lubricant.

Another way to reduce friction between two parts is to superimpose micro-scale vibration to one of the parts. This can be sinusoidal vibration as used in ultrasound-assisted cutting or vibration noise, known as dither.

Energy of Friction

According to the law of conservation of energy, no energy is destroyed due to friction, though it may be lost to the system of concern. Energy is transformed from other forms into heat. A sliding hockey puck comes to rest because friction converts its kinetic energy into heat. Since heat quickly dissipates, many early philosophers, including Aristotle, wrongly concluded that moving objects lose energy without a driving force.

When an object is pushed along a surface, the energy converted to heat is given by:

$$E_{th} = \mu_k \int F_n(x)dx$$

where

F_n is the normal force,

μ_k is the coefficient of kinetic friction,

x is the coordinate along which the object transverses.

Energy lost to a system as a result of friction is a classic example of thermodynamic irreversibility.

Work of Friction

In the reference frame of the interface between two surfaces, static friction does *no* work, because there is never displacement between the surfaces. In the same reference frame, kinetic friction is always in the direction opposite the motion, and does *negative* work. However, friction can do *positive* work in certain frames of reference. One can see this by placing a heavy box on a rug, then pulling on the rug quickly. In this case, the box slides backwards relative to the rug, but moves forward relative to the frame of reference in which the floor is stationary. Thus, the kinetic friction between the box and rug accelerates the box in the same direction that the box moves, doing *positive* work.

The work done by friction can translate into deformation, wear, and heat that can affect the contact surface properties (even the coefficient of friction between the surfaces). This can be beneficial as in polishing. The work of friction is used to mix and join materials such as in the process of friction welding. Excessive erosion or wear of mating surfaces occur when work due frictional forces rise to unacceptable levels. Harder corrosion particles caught between mating surfaces (fretting) exacerbates wear of frictional forces. Bearing seizure or failure may result from excessive wear due to work of friction. As surfaces are worn by work due to friction, fit and surface finish of an object may degrade until it no longer functions properly.

Screw Jack

A jackscrew is a type of jack which is operated by turning a leadscrew. In the form of a screw jack it is commonly used to lift heavy weights such as the foundations of houses, or large vehicles.

Advantages

An advantage of jackscrews over some other types of jack is that they are *self-locking*, which means when the rotational force on the screw is removed, it will remain motionless where it was left and will not rotate backwards, regardless of how much load it is supporting. This makes them inherently safer than hydraulic jacks, for example, which will move backwards under load if the force on the hydraulic actuator is accidentally released.

Mechanical Advantage

The mechanical advantage of a screw jack, the ratio of the force the jack exerts on the load to the input force on the lever, ignoring friction, is

$$\frac{F_{\text{load}}}{F_{\text{in}}} = \frac{2\pi r}{l}$$

where

F_{load} is the force the jack exerts on the load

F_{in} is the rotational force exerted on the handle of the jack

r is the length of the jack handle, from the screw axis to where the force is applied

$l$is the lead of the screw.

However, most screw jacks have large amounts of friction which increase the input force necessary, so the actual mechanical advantage is often only 30% to 50% of this figure.

Applications

A jackscrew's threads must support heavy loads. In the most heavy-duty applications, such as screw jacks, a square thread or buttress thread is used, because it has the lowest friction. In other application such as actuators, an Acme thread is used, although it has higher friction. The large area of sliding contact between the screw threads means jackscrews have high friction and low efficiency as power transmission linkages, around 30%–50%. So they are not often used for continuous transmission of high power, but more often in intermittent positioning applications.

The Ball screw is a more advanced type of leadscrew that uses a recirculating-ball nut to minimize friction and prolong the life of the screw threads. The thread profile of such screws is approximately semicircular (commonly a "gothic arch" profile) to properly mate with the bearing balls. The disadvantage to this type of screw is that it is not self-locking.

Jackscrews form vital components in equipment. For instance, the failure of a jackscrew on a McDonnell Douglas MD80 airliner due to a lack of grease resulted in the crash of Alaska Airlines Flight 261 off the coast of California in 2000.

The jackscrew figured prominently in the classic novel *Robinson Crusoe*. It was also featured in a recent History Channel programme as *the* saving tool of the Pilgrims' voyage – the main crossbeam, a key structural component of their small ship, cracked during a severe storm. A farmer's jackscrew secured the damage until landfall.

In Electronic Connectors

The term *jackscrew* is also used for the removable screws that hold D-subminiature electrical connectors together. These screws draw the two connector halves together and hold them mated, and when unscrewed allow the connector halves to be taken apart. These small jackscrews may have ordinary screw heads or extended heads (also making them thumbscrews) that allow the user's fingers to turn the jackscrew. Furthermore, the head sometimes has an internal female thread, with the male externally threaded screw shaft extending from that. The threaded-head type can be used to panel-mount one connector and provide a means to attach the mating connector to the first connector.

Differential Screw Jack

A differential screw is a mechanism used for making small, precise adjustments to the spacing between two objects (such as in focusing a microscope, moving the calipers of a micrometer, or positioning optics). A differential screw uses a spindle with two screw threads of differing leads (aka thread pitch), and possibly opposite handedness, on which two nuts move. As the spindle rotates, the space between the nuts changes based on the difference between the threads. These mechanisms allow extremely small adjustments using commonly available screws. A differential screw mechanism using two nuts incurs higher friction and therefore requires more torque to turn than a simple, single lead screw with an equivalent pitch.

Examples

Many differential screw configurations are possible. The micrometer adjuster pictured uses a nut sleeve with different inner and outer thread pitches to connect a screw on the adjusting rod end with threads inside the main barrel; as the thimble rotates the nut sleeve, the rod and barrel move relative to each other based on the differential between the threads.

Another arrangement holds the two "nuts" co-axially in a single fixture and has two separate screws with slightly different pitches (threads per inch) entering from opposite ends. The "heads" of the screws are fixed to the two objects whose spacing is to be adjusted. Each rotation of the fixture holding the nuts moves one screw into its nut by a small amount and moves the other screw out of its nut by a slightly larger amount. The total spacing between the screws, and thus the objects, will be slightly changed based on the difference in travel between the two screws.

More arrangements are possible. Two nuts can be fixed to each of to objects to be adjusted and the two screw heads attached to each other in the middle. The combined screws would be turned to adjust the spacing in that case.

Bibliography

Armstrong-Hélouvry, Brian: *Control of Machines with Friction*, Springer, USA, 1991.

Beer, Ferdinand P.; E. Russel Johnston, Jr.: *Vector Mechanics for Engineers*, McGraw-Hill, NY, 1996.

Brown, R., Paton, R. and T. Porter: *Computation in Cells and Tissues - Perspectives and Tools of Thought*, Springer Verlag, 2004.

Brush, S.: *The Kind of Motion That We Call Heat*, North-Holland, Amsterdam, 1976.

Buckel, W.: *Superconductivity: Fundamentals and Applications,* VCH Publishers, Inc., 1991.

Coleman, Michael M.: *Fundamentals of Polymer Science: An Introductory Text*, Technomic Pub. Co., Lancaster, PA., 1997.

Dahl, P. F.: *Superconductivity: Its Historical Roots and Development from Mercury to the Ceramic Oxides,* American Institute of Physics, 1992.

David Chandler: *Introduction to Modern Statistical Mechanics,* Oxford University Press, Oxford, 1987.

David Chandler: *Introduction to Modern Statistical Mechanics,* Oxford University Press, Oxford, 1987.

Dowson, Duncan: *History of Tribology,* Professional Engineering Publishing, UK, 1997.

Faulkner, L.R.: *Electrochemical Methods: Fundamentals and Applications*, John Wiley & Sons, New York, 2005.

Freyd, P.: *Functor Theory,* Princeton University, Princeton, New Jersey, 1960.

Gibbs, J.: *Elementary Principles in Statistical Mechanics*, Dover, New York, 1960.

Hibbeler, R. C.: *Engineering Mechanics,* Pearson, Prentice Hall, 2007.

Ibach H. and H. Lýÿth: *Solid-State Physics: an Introduction to Principles of Materials Science,* Springer, Berlin; New York, 1996.

Jacobs, B.: *Categorical Logic and Type Theory*, Amsterdam, North Holland, 1999.

Lambek, J. & Scott, P.J.: *Introduction to Higher Order Categorical Logic*, Cambridge University Press, Cambridge, 1986.

Mark W. Zemansky: *Heat and Thermodynamics*, McGraw-Hill, Delhi, 1968.

Marquis, J.-P.: *Russell and Analytic Philosophy*, University of Toronto Press, Toronto, 1993.

Meriam, J. L.; L. G. Kraige: *Engineering Mechanics*, John Wiley & Sons., NY, 2002.

Meriam, James L.; Kraige, L. Glenn and Palm, William John: *Engineering Mechanics: Statics*, Wiley and Sons, London, 2002.

Michael C.: *Time's Arrow: The Originis of Thermodynamic Behaviour*. Berlin Heidelberg New York: Springer, 1992.

Nosonovsky, Michael: *Friction-Induced Vibrations and Self-Organization: Mechanics and Non-Equilibrium Thermodynamics of Sliding Contact*, CRC Press, US, 2013.

Penrose, R.: *Shadows of the Mind*, Oxford University Press, Oxford, 1994.

Rob Jenkins and C.K. Jain: *Advances in Soil-Borne Plant Diseases*, Oxford Book Company, Delhi, 2010.

Ruina, Andy; Rudra Pratap: *Introduction to Statics and Dynamics*, Oxford University Press, Oxford, 2002.

Samuel W. Johnson: *Crops Feed from Air and Soil*, Reprint Pub, Delhi, 2005.

Sheppard, Sheri; Tongue, Benson H. and Anagnos, Thalia: *Statics: Analysis and Design of Systems in Equilibrium*, Wiley and Sons, London, 2005.

Singh A K: *Environment and Water Resources Management*, Adhyayan, Delhi, 2006.

Somani, L.L.: *Diagnosis and Improvement of Acid Soils*, Agrotech Pub, Delhi, 2009.

Soutas-Little, Robert W.; Inman, Balint: *Engineering Mechanics*, Thomson Press, Haryana, 2008.

Weibel-Mihalas, B.: *Foundations of Radiation Hydrodynamics*, Oxford University Press, New York, 1984.

Index

M

N

O

P

Q

R

S

T

V

W

❑❑❑